Vintage Reports II: International Media Coverage of the Ningxia Wine Scene

年份的足迹 II：国际媒体报道宁夏葡萄酒

洪润宇（Runyu Hong）
[加] Jim Boyce（吉姆·博伊斯）　编　译
马会勤（Huiqin Ma）

中国农业大学出版社
China Agricultural University Press

内 容 简 介

本书收录了从 2015 年到 2021 年 3 月国际主流媒体和葡萄酒专业媒体对宁夏葡萄酒产业的报道，是宁夏葡萄酒依靠优良品质获得国际葡萄酒产业和知名专家赞誉的扎实记录。文章和报道中不乏褒扬赞美之词，同时专家也指出了产业发展中面临的困难与挑战。如书名所展示的，《年份的足迹Ⅱ：国际媒体报道宁夏葡萄酒》记录下了宁夏葡萄酒产业发展中每一年的步伐，和葡萄酒的年份一样，这是不断克服困难和挑战、追求卓越的过程。

图书在版编目（CIP）数据

年份的足迹：国际媒体报道宁夏葡萄酒 .Ⅱ：汉、英 / 洪润宇，（加）吉姆•博伊斯（Jim Boyce），马会勤编译 .--北京：中国农业大学出版社，2021.7
ISBN 978-7-5655-2579-7

Ⅰ.①年… Ⅱ.①洪… ②吉… ③马… Ⅲ.①葡萄酒 – 介绍 – 宁夏 – 汉、英②新闻报道 – 作品集 – 世界 – 现代 – 汉、英　Ⅳ.① TS262.6 ② I15

中国版本图书馆 CIP 数据核字（2021）第 147147 号

书　　名	年份的足迹Ⅱ：国际媒体报道宁夏葡萄酒
作　　者	洪润宇　Jim Boyce　马会勤　编译

策划编辑	梁爱荣	**责任编辑**	何美文　梁爱荣
封面设计	郑　川		
出版发行	中国农业大学出版社		
社　　址	北京市海淀区圆明园西路 2 号	**邮政编码**	100193
电　　话	发行部 010-62733489, 1190	**读者服务部**	010-62732336
	编辑部 010-62732617, 2618	**出 版 部**	010-62733440
网　　址	http://www.caupress.cn	**E-mail**	cbsszs@cau.edu.cn
经　　销	新华书店		
印　　刷	北京鑫丰华彩印有限公司		
版　　次	2021 年 8 月第 1 版　2021 年 8 月第 1 次印刷		
规　　格	787 mm×1 092 mm　16 开本　16.5 印张　250 千字		
定　　价	86.00 元		

图书如有质量问题本社发行部负责调换

编译校人员名单

编　译

洪润宇　Jim Boyce　马会勤

参译和校对

李昭航　扈朝阳　翟雨佳　郝瑞龙

本书的出版得到下列机构的支持，在此特别致谢！

宁夏贺兰山东麓葡萄产业园区管理委员会
宁夏贺兰山东麓葡萄酒教育学院

序 preface

　　我们用这本书致敬时间，致敬那些在每一个酿酒年份里应该被我们记住的人和事。

　　宁夏的葡萄酒是如何开始被人们知道和记住的？仅在20年前，宁夏还只是给大品牌做葡萄种植基地的新发展区域。作为行业发展的参与者和观察者，我们见证了宁夏在10年左右的时间里，成为国际媒体和国内外专家谈论中国葡萄酒时最常被提及的产区。当人们谈到中国最优质的葡萄酒群体时，他们说：宁夏。

　　没有任何成功来得容易，葡萄酒在某种意义上说是时间的艺术，在一个个年份的打磨中，一个产区逐渐知道了哪些品种更适合在当地种植，要用什么样的砧木、架式和种植密度，什么样的酿造方法能更好地展示这块土地的奉献。葡萄酒还需要让媒体和消费者知道并了解它们的性格、特点，并为人们所喜欢。

　　这个世界上总有一些人胸怀更高的追求，更肯付出，更谦逊，更宽容，这是有追求的生命的美好。因为这些美好的生命、聪明的头脑、勇敢的性格，我们看到探索、创造，看到坚守、成功，看到宁夏葡萄酒产区在一个个年份中脱颖而出。

　　宁夏葡萄酒的成功是这片土地的成功，也是人的成功。从宁夏葡萄酒产业的领导者，到酿酒师，从企业老板，到在葡萄园工作的搬迁移民，他们的勤劳和智慧让这片曾经的荒芜之地，变成全世界最值得到访的葡萄酒之乡。他们为这个产业的付出，值得我们永远记在心里。

  本书收录了从 2015 年到 2021 年 3 月国际媒体对宁夏葡萄酒的报道，我们也感谢这些文章的作者，向世界讲述和传播了宁夏葡萄酒的故事。本书按年度进行编辑，如同一个个葡萄酒年份，当积累了足够长的时间，一次垂直品鉴可以帮助我们回顾产区在时间中的进步，那些我们永远不会忘记的人和故事会再次被提起，重温美好。

  因为这一切，无论艳阳还是星空之下，我都愿以一杯美酒，致敬你的奉献！

2021 年 4 月

Preface

Ten winters, ten springs, ten summers, ten falls. These are but a short while for a wine region, where evolution is at times measured in centuries. Even so, in that brief span, in those ten vintages, Ningxia has risen from curiosity to contender in the wine world.

A decade or so ago, Ningxia was largely unknown as a wine region and Chinese wines generally inspired scorn. Season after season of hard work and hard lessons saw Ningxia's wine industry improve and expand, and earn nearly a thousand contest medals paired with kudos from critics. Now that industry has turned its attention to finding similar success with consumers. In fact, development is so fast that we see sub-regional wine associations confidently heading beyond Ningxia's borders to promote their brands.

That's quite a wine story. But it's even more meaningful. It's also a story about land reclamation—turning what some saw as wasteland into vineyards. About poverty reduction—raising people from subsistence living to greater levels of comfort and security. And about pride—providing a mission for everyone from policy makers to grape pickers, from wine makers to wine marketers, to put their region in the national spotlight.

Incredibly, the growth rate is set to accelerate this year, with plans for more vineyards, wines, tourists and jobs. No doubt there are struggles ahead: local spirits and beers still dominate the market; many consumers remain unaware of Ningxia wine; and the industry is still learning what grapes, techniques et all work best. But these are "good problems", only possible because of the incredible progress made thus far.

For now, hope and confidence abound. The continued rise in wine quality, the growing interest in local brands, notably among young consumers, and unexpected boosts like a well-received TV drama this year called Mountain Sea Love that covers two decades of poverty alleviation and features the fruits of that labor—Ningxia wine!—in the final episode, are among the inspirations.

This book looks at that Ningxia wine story through the eyes of dozens of media articles. Just as each winery makes a unique wine, each report gives a "taste" of the tale unfolding in this region, of a plot that developed at lighting speed and shows no signs of slowing.

April, 2021

2015

迈宁格国际葡萄酒商业：走进宁夏 …………………………………………… 03
纽约时报：中国酿酒师寻找自己的纳帕 ………………………………………… 09
霍克湾今日报：湾区酿酒师期待在中国留名 …………………………………… 11
醇鉴：宁夏葡萄酒摘金 …………………………………………………………… 13
Wine-Searcher：破解中国葡萄酒之谜 ………………………………………… 14
华尔街日报：法国让位，中国来了 ……………………………………………… 15
美食家：马蒂亚斯·瑞格在中国出名 …………………………………………… 16
印度葡萄酒学院：两名印度人参加宁夏葡萄酒挑战赛 ………………………… 17

2016

CBS 新闻：中国把荒漠变成葡萄酒产区的豪赌 ………………………………… 21
电讯报：中国葡萄酒的红色曙光 ………………………………………………… 22
醇鉴：中国开展太空葡萄试验 …………………………………………………… 23
世界报：中国宁夏的红色黄金 …………………………………………………… 24
赫曼努斯时报：南非酿酒师在宁夏大展技艺 …………………………………… 25

2017

马尔堡快报：市长访问中国签订姐妹城市协议 ………………………………… 29
买手杂志：中国葡萄酒质量的攀登之路 ………………………………………… 31

01

醇鉴：生物动力法在宁夏 …………………………………………………… 34
北京人：品鉴中国葡萄酒的完美入门清单 ………………………………… 38
中国日报亚洲版：各国酿酒师参加宁夏葡萄酒挑战赛 …………………… 39

2018

WineLand：超越纳帕，宁夏驾到 …………………………………………… 43
中国日报：铁路上的葡萄酒 ………………………………………………… 47
美食家：穿越中东的葡萄酒之路——中国最优葡萄酒 …………………… 51
罗宾逊网站：宁夏的青春锋芒 ……………………………………………… 55
中国日报美国版：宁夏葡萄酒惊艳联合国 ………………………………… 62
饮料商务：宁夏葡萄酒列车鸣笛发车 ……………………………………… 65
Just-Drinks：为什么世界葡萄酒商不能在中国为所欲为 ………………… 66

2019

Wine-Searcher：中国葡萄酒破壳而出 ……………………………………… 69
饮酒杂志：龙的秘密——宁夏如何将中国葡萄酒标上世界地图 ………… 73
印度葡萄酒学院：第三届世界葡萄酒博览会彰显宁夏的中国顶级
　　葡萄酒产区地位 ………………………………………………………… 79
贝丹和德梭：贺兰红，中国葡萄酒新品牌 ………………………………… 82
葡萄圈：夏桐推出其中国首款起泡红葡萄酒 ……………………………… 84

2020

华盛顿邮报：引领中国葡萄酒革命的女性 ………………………………… 87
饮料商务：宁夏产区继续成长 ……………………………………………… 88
人民日报：葡萄酒产业使宁夏的贫瘠土地变得繁荣 ……………………… 90
北京评论：东西部对口合作促进贫困地区发展 …………………………… 92
新华社：中国最重要葡萄酒产区的远大梦想 ……………………………… 96

南华早报：为何中国的生物动力葡萄酒既能畅销国内，又能受到
　　欧洲和日本鉴赏家的喜爱 ································· 99
环球邮报：首次中国葡萄酒品鉴展现巨大潜力 ······················ 101
福布斯：宁夏葡萄酒亮相墨西哥城 ································ 102
中国日报：宁夏入选全球十大最具发展潜力的葡萄酒旅游目的地 ······ 103
英国广播公司新闻频道：中国消费者对国产葡萄酒日渐喜爱 ·········· 104

2021

葡萄酒评论：宁夏，征服沙漠的葡萄酒 ···························· 107
爱尔兰时报：走进东方巨龙——包含西方经验的中国葡萄酒 ·········· 109
詹姆斯·萨克林网站：2020宁夏报告——快钱 vs 值得陈年的质量 ····· 110

2015

Meininger's Wine Business International: Inside Ningxia ················ 123
The New York Times: China's Winemakers Seek Their Own Napa Valley ··· 130
Hawke's Bay Today: Bay Winemaker Aims to Make His Mark in China ······ 132
Decanter: DAWA 2015: Ningxia Chinese Wines Scoop Gold Medals ·········· 134
Wine-Searcher: Solving the Chinese Wine Puzzle ························ 135
The Wall Street Journal: Move Over, France—Here Comes China ·········· 136
Gourmetwelten: Mathias Regner Big in China ···························· 137
Indian Wine Academy: Two Indians at Ningxia Wine Challenge ············ 138

2016

CBS News: China Makes Big Bet on Turning Desert into Wine Region ······ 141
The Telegraph: Red Dawn for Chinese Wine ······························ 142
Decanter: China Grows Wine in Space to Beat Harsh Climate ············· 143
Le Monde: En Chine, l' or Rouge du Ningxia ···························· 144
Hermanus Times: Local Winemaker Impresses Chinese Connoisseurs ········ 145

2017

Marlborough Express: Mayor Takes Trip to China to Sign Sister-city Agreement ·················· 149

The Buyer: On the Road: How China is Stepping up the Quality of Its Wines 151

Decanter: Going Biodynamic in Ningxia ·················· 155

The Beijinger: Annual Grape Wall Challenge Gives You A Perfect Starter List for Chinese Wines to Try ·················· 159

China Daily Asia: Winemakers Rise to Ningxia Challenge ·················· 160

2018

WineLand: Move Over Napa, Here Comes Ningxia ·················· 163

China Daily: Wine Down the Track ·················· 168

Gourmetwelten Das Genussportal: Auf der Wein-Route durch das Reich der Mitte: Beste Weine Chinas ·················· 172

jancisrobinson.com: Young Guns of Ningxia ·················· 177

China Daily USA: Ningxia Wines Wow UN Diplomats ·················· 185

The Drinks Business: China Launches Ningxia Wine Train from Beijing ·················· 188

Just-Drinks: Why the World's Wine Producers Won't Have It All Their Own Way in China ·················· 189

2019

Wine-Searcher: China's Wine Egg Hatches ·················· 193

Imbibe Magazine UK: Enter the Dragon: How Ningxia Is Putting Chinese Wine on the Map ·················· 197

Indian Wine Academy: 3RD Edition of BRWSC Highlights Ningxia as Top Wine Region of China ·················· 204

Bettane + Desseauve: Helan Hong, Nouvelle Marque de Vin Chinoise ·················· 208

Vitisphere: Chandon Lance Son Premier Vin Rouge Pétillant de Chine ……… 210

2020

The Washington Post: These Are the Women Leading China's Wine
　　Revolution ……………………………………………………… 213
The Drinks Business: Ningxia Region Continues to Grow ……………… 214
People's Daily: Grape Wine Industry Turns Ningxia's Barren Land into
　　Prosperity ……………………………………………………… 216
Beijing Review: East-west Cooperation Program Facilitates Development of
　　Poverty-stricken Region ………………………………………… 219
Xinhua News: China's Major Winemaking Region Dreams Big ………… 223
South China Morning Post: Why China's Biodynamic Wines are Gaining
　　a Cult Following ………………………………………………… 226
The Globe and Mail: The First Real Taste of Chinese Wine Shows
　　Tremendous Potential …………………………………………… 227
Forbes: From Competition to Cantina: Concours Mondial de Bruxelles
　　Opens Its First Wine Bar in Mexico City ……………………… 228
China Daily: Ningxia Named Among World's 10 Most Promising
　　Wine Destinations ………………………………………………… 229
BBC News: China's Drinkers Develop Taste for Home-grown Wines ……… 230

2021

La Revue du Vin de France: Ningxia, le Vin à la Conquête du Désert ……… 233
The Irish Times: Enter the Dragon: Chinese Wines Embodying Lessons
　　from the West …………………………………………………… 235
jamessuckling.com: 2020 Ningxia Report: Fast Money vs Age-worthy
　　Quality …………………………………………………………… 236

2015

迈宁格国际葡萄酒商业
走进宁夏

原文链接：
wine-business-international.com/wine/general/ inside-ningxia

作者：吉姆·博伊斯
2015年9—10月刊

吉姆·博伊斯（Jim Boyce）解释说，中国政府正在大力支持葡萄酒产业的发展，将其视为遏制荒漠化扩张的一项措施。这使宁夏得以快速、充满活力地发展。

在中国西北部一抹黄色的地形图上，宁夏的葡萄酒产区是一道绿色。这个最著名的产区坐落在贺兰山和黄河之间，贺兰山阻挡了来自西边的干冷寒风，黄河从东边为葡萄酒产区提供生死攸关的水源。

即使在5年之前，如果你提到宁夏，大多数人也会回答"嗯？"而现在，酿酒师、葡萄栽培师、顾问和投资者源源不断地涌入这里。根据该地区与葡萄酒行业最相关的三名政府官员郝林海、李学明和曹凯龙的一份2015年的报告，酿酒葡萄的种植面积从2005年的不到3 000公顷，10年间增加到近40 000公顷。鉴于政府的目标是到2020年达到66 000公顷，还有很多工作要做。曹凯龙说，现有的酒庄数量为72个，还有58个在建和计划建设的，并且计划增加更多的小型精品酒庄。

一个地区的崛起

大约30年前，宁夏的葡萄酒产业向位于中国东部的山东省和河北省的张裕和长城等生产商提供散装的原酒。到了20世纪90年代，当时政府开始对开垦荒地和催生本地增值品牌产生兴趣。这项努力随着前政府官员容健于2001年成立宁夏葡萄产业协会而显著加强。而最大的变化发生在2012年，与政府的"十二五"规划相一致，该规划旨在进一步发展葡萄酒产区，特别是在中国西北部地区，该规划还将有助于遏制荒漠化的发展。正如2012年世界银行在批准一笔8 000万美元的贷款以帮助宁夏抗击沙漠的侵蚀时所阐述的那样，荒漠化影响了该地区近300万公顷的土地，这部分面积占该地区总面积的57%。

2012年曹凯龙组建了葡萄花卉产业局，后更名为葡萄产业发展局，以帮助推动变革。这项举措回报丰富且快速，因为宁夏不仅有良好的葡萄种植条件，而且比东部葡萄酒生产省拥有更低的一般人工成本和土地成本。能够租用大片土地并种植自己的葡萄，而不必依赖从农民那里购买质量不均的葡萄果实，这一点吸引了投资者们入驻该产区。

宁夏在与更广阔的葡萄酒世界的联系方面也取得了卓越的进步：它成为中国第一个国际葡萄与葡萄酒组织（OIV）省级观察员（沿海省份山东的烟台市是市级的OIV观察员），并派代表团参加了OIV年度大会。实地调研小组观察梳理了新西兰、澳大利亚、法国和美国等国家的最佳葡萄酒实践经验。宁夏还在银川举办了葡萄酒大会、葡萄酒节和贸易博览会，其中包括SiteVintech（包括酿酒葡萄在内的国际水果种植贸易博览会），并为葡萄苗的进口和建立苗圃提供了便利。宁夏还有一些其他项目，包括建立列级酒庄体系,以及组织一系列为期两年的名为"宁夏酿酒师挑战赛"的项目。本届宁夏酿酒师挑战赛设立了 11.2 万美元的现金奖励,将当地生产者与 48 个国际酿酒师配对,以促进文化和葡萄栽培的交流。

可以肯定地说,很少有葡萄酒产区在这么短的时间内有如此多的活动。去年开业的宁夏国际葡萄酒贸易博览中心具有深远意义,可以说是这些努力的象征。联合会主席郝林海说,该中心的目的是"促进与 OIV 成员的合作,建立一个永久性的葡萄酒博物馆,并推广贺兰山东麓葡萄酒"。

葡萄酒品质

精选的宁夏葡萄酒在国内和国际比赛中均获得了奖牌,并获得了米歇尔·贝丹(Michel Bettane)、切里·德梭(Thierry Desseauve)、杰里米·奥利弗(Jeremy Oliver)、安德鲁·杰佛德(Andrew Jefford)和凯伦·麦克尼尔(Karen MacNeil)等评论家的好评。2012 年,葡萄酒大师杰西丝·罗宾逊(Jancis Robinson)在宁夏品尝了 40 款当地葡萄酒,她将 5 款评为极好,而仅有 6 款被评为是不适于商业的,且它们的主要缺点是氧化——她指出氧化是相对容易解决的缺陷。贺兰晴雪酒庄出品的一款葡萄酒在 2011 年荣获醇鉴国际大奖,并被作为全球头条新闻报道。虽然人们普遍认为宁夏呈现以波尔多风格的赤霞珠为主导的风格,缺乏多样性,但至少该地区已经建立了它能够生产优质葡萄酒的声誉。

宁夏的葡萄酒产区也很特别,它非常紧凑,主要的酒庄沿着贺兰山山脚星散分布,距省会银川不到一小时的车程。与之相比,山西省的怡园酒庄(Grace Vineyard)可以说是中国最好的酒庄,但它在当地就是孤单的存在。或对比位于遥远西北的地域辽阔的新疆,那里酒庄之间通常相隔数小时的车程。

位于北京的中国农业大学的马会勤教授今年开始正式担任该局的职务,并表示:"酒庄之间距离近有利于交流,对酿酒师、葡萄栽培师和管理人员都是如此,这对发展中的产业至关重要。"马会勤教授还提到"旅游效率"是一个优势。她说:"人们可以在半天之内参观 3 个独特的酒庄,从夏桐这样的起泡酒酒庄,到像迦南美地这样生产偏德国风格的葡萄酒的酒庄。"

宁夏的酒庄涵盖了多种商业模式。其中包括大型葡萄酒生产商在当地的分支机构：生产长城葡萄酒的中粮集团（COFCO）拥有一个占地1 500公顷的酒庄，名为云漠，而张裕与奥地利酿酒师伦茨·摩塞尔（Lenz Moser）联手建立了具有卢瓦尔河谷风格的城堡和酒庄，还设有博物馆。外资项目中突出的是饮料巨头如保乐力加（Pernod Ricard）和路易·威登－酩悦·轩尼诗（LVMH），保乐力加拥有其贺兰山品牌，而路易·威登－酩悦·轩尼诗拥有夏桐酒庄，酒庄去年生产了第一个年份的商业起泡酒。其他知名度稍低的外资项目还包括德隆，由泰国德盛集团（Daysun）投资，它拥有超过6 000公顷的葡萄园。

这里还包括银色高地和贺兰晴雪这样的酒庄，银色高地以2007年第一个年份的葡萄酒将宁夏标注在世界葡萄酒地图上，而贺兰晴雪如前所述在醇鉴大赛上获得国际大奖。加上其他数十家酒庄，从老将西夏王酒庄和广夏酒庄，到生产冰酒的森淼兰月谷酒庄，在多样性方面，中国的任何其他地区都无法与宁夏比肩。

这种多样化会进一步增长，通过集中精力建立更多的小酒庄，尤其是那些年产15万～20万瓶的酒庄，可以更好地发掘宁夏产区的潜力。

曹凯龙在接受《中国日报》的采访时说："这就像书法，每个人都有自己的风格。"

挑战

该地区面临的最大挑战可能恰恰是其快速的成功。与5年前该地区首次亮相时相比，现在人们的期望更高，葡萄酒评论家的宽容度也更低。葡萄酒评论家会承认宁夏可以酿造出优质的葡萄酒，但对其葡萄酒能否达到卓越的水准持观望态度。法国作家贝丹（Bettane）和德梭（Desseauve）说，沙质土壤占主导地位，缺乏黏土，尤其是深层的土壤，将使这里产生柔顺的果香型葡萄酒，但葡萄酒的复杂度可能不会特别高。在某些葡萄酒中，尤其是在蛇龙珠（当地人对佳美娜的称谓）中，有一种不太友好的生青植物感。这可能是葡萄栽培方式不合理造成的，如果确实是这样的话，这个问题是可以被解决的。

北京农学院教授、贺兰晴雪酒庄的首席顾问李德美一直强调，宁夏要想赶上波尔多，还有很长的一段路要走。

这段路有多长呢？他的门徒酿酒师张静，用10年为单位来丈量。她说："我们需要酿造出无缺陷的葡萄酒，不是一个酒庄达到这样的水准，而是整个地区。经过30～50年，我们可以慢慢地从风土中提炼出某种特征，并且可以说这就是宁夏葡萄酒。"

张静在回应有关沙质土壤的评论时指出，宁夏的许多土壤类型还有待进一步的探索。银色高地将这一点付诸实践。有人说银色高地算得上中国第一家"车库"酒厂，酿酒师高源（Emma Gao）在谈到她家在一个与其他葡萄园不相邻的多石的地区建立新葡萄园时说："通过多年在宁夏酿酒的实践，我父亲和我选择了不太一样的区域种植我们家的葡萄。"

一项不能解决或改善，而不得不与之共存的条件是气候。宁夏的冬季不仅寒冷干燥，而且该地区也遭受沙暴和尘暴的侵袭，必须在秋天掩埋葡萄藤，并在春天出土。虽然从短期看埋土可以保护葡萄藤，但这样做的代价是会损失一些葡萄藤，同时还会增加成本。缺水是另一个重要难题，尤其是当葡萄酒行业的规模不断扩大，存在葡萄藤与其他农作物竞争水源的问题。今年，宁夏政府宣布启动一项7 600万美元的应用以色列滴灌技术的项目，该项目将灌溉80个酒庄共计9 200公顷的葡萄园。

随着年轻人向城市的转移，劳动力变得稀缺。郝林海、李学明和曹凯龙在2015年的报告中说，政府已经将农民从宁夏的贫困山区转移到了酿酒产区，以帮助缓解劳动力短缺。他们写道："将搬迁的农民培养成酿酒葡萄种植者，从而使家庭生活水平和收入得到了显著的改善。"尽管这项工作的可持续性以及进一步找寻新的劳动力的前景还有待进一步观察。不出意料的是，新的酒庄很可能采用机械化的葡萄栽培。

虽然曾经有人认为由于宁夏地区气候干燥葡萄没有病害，但细致的研究发现卷叶病毒病和葡萄干部病害的问题比以前预想的要多。这在2012年变得尤为明显，异常的潮湿天气导致霉菌泛滥，并迫使人们深入研究葡萄园病害。

最后，按照当地标准，葡萄酒的价格往往高得令人望而却步，被葡萄酒评论家品尝过的酒款通常被定价在50美元或更高。酿造出国际葡萄酒评论家认为没有缺陷的优质葡萄酒是一回事，以能够与大量优质、廉价的进口葡萄酒竞争的价格销售它们是另一回事。如果宁夏能够将自己发展成为类似美国纳帕谷那样的葡萄酒产区，这样的价格也许行得通。考虑到包括郝林海在内的地方政府官员对葡萄酒质量的重视（郝林海曾担任过宁夏回族自治区人民政府副主席和省会银川市市长），成为"中国的纳帕谷"无疑是宁夏的目标。

尽管面临所有这些挑战，宁夏目标高远、进步神速。过早采摘葡萄或默认使用新的法国橡木桶之类的做法已经迅速转变。"几年前，我们还在担心葡萄能否具备足够的糖分，但是现在这很容易。"酿酒师张静说，"我们现在的关注点是酸度和果实成熟度。"一些生产商也逐渐做出不同于波尔多混酿的新尝试，出现了更多的雷司令、黑比诺、西拉、马瑟兰，甚至出现了白比诺。人们致力于研究哪种葡萄表现最好，包括不需要埋土的葡萄品种，探索从灌溉技术到叶幕管理的所有方面。但是，宁夏最激励人心的也许是该地区许多酿酒师之间的合作。他们经常会面，品尝彼此的作品，并谈论如何使整个地区的葡萄酒更好，甚至每个人都在考虑如何赢得下一枚金牌。

纽约时报
中国酿酒师寻找自己的纳帕

原文链接：

nytimes.com/2015/11/08/business/international/chinas-winemakers-seek-to-grow-their-own-napa-valley.html

作者：简·萨森
2015 年 11 月 7 日

  从精品酒庄模式得到启发，宁夏有着成为"中国的纳帕谷"的野心。本地酿酒师已赢得了多个著名奖项，同时该地区正在计划将葡萄园的面积翻倍，打造一个葡萄酒旅游中心。海外投资者们同样注意到了这里。法国酩悦香槟的酿造商在宁夏生产起泡葡萄酒，而烈酒巨头保乐力加也在花费重金对其在本地的酒庄进行现代化改造。

  "人们知道纳帕酿造美国最好的葡萄酒，波尔多酿造法国最好的葡萄酒，"管理宁夏葡萄酒产业的地方政府官员郝林海说，"而提到中国葡萄酒，我们希望人们想到宁夏"。

英国葡萄酒商人史蒂芬·史普瑞尔（Steven Spurrier）组织过1976年那场著名的"巴黎评判"，在那场葡萄酒盲品大赛上，加利福尼亚葡萄酒战胜了法国酒，一举震惊了世界葡萄酒行业。他说："他们有来自全世界的钱，有全世界最大的野心，还把最好的顾问都请了来，中国人必然会酿造出越来越好的葡萄酒。"（节选）

霍克湾今日报
湾区酿酒师期待在中国留名

原文链接：

nzherald.co.nz/hawkes-bay-today/news/ bay-winemaker-aims-to-make-his-mark-in-china/SQRNHJ3G4PTH 6GGC536 QMIKOH4/

作者：艾米·沙克斯

2015年9月4日

从新西兰霍克湾的蒂阿瓦昂加（Te Awanga）到位于中国西北部的宁夏路途遥远，但如果有酒可酿，距离和地点都不是问题。

对丽景庄园酒庄（Clearview Estate Winery）的新任酿酒师马特·柯比来说，如果两年后他在中国酿造的葡萄酒能得到评委团的赞许，这将是一件值得骄傲的事，更不用说他还能赚到不错的一笔小钱。

柯比先生和来自世界各地的其他59位酿酒师一起，将参加由贺兰山东麓国际葡萄酒联合会在宁夏葡萄产业发展局支持下主办的第二届宁夏葡萄酒挑战赛，争夺逾20万美元的奖金。

现年33岁的柯比先生于今年年初加入丽景，他是全球140多位报名者中仅有的7名新西兰酿酒师之一，也是唯一来自霍克湾的酿酒师。

他们的任务是用中国最有前途的葡萄酒产区的葡萄酿造出一款获奖葡萄酒。

柯比先生将于9月18日飞往北京以西1 100千米的宁夏，在那里用15天时间进行葡萄的筛选和初步发酵。

在为期两年的挑战赛期间，他最多将返回宁夏6次，每次停留5～15天，管理他酿造的葡萄酒，直到2017年秋天，届时葡萄酒将接受品鉴评比。

他的所有差旅费用都由活动组织者承担。

"参加这次挑战赛令我倍感兴奋。"他说。

"中国的葡萄酒行业发展如此之快，亲眼目睹和亲身体验这一切将是一件非常有趣的事情，因为这一切都是全新的，所有的技术都是最新的，这真是令人兴奋。"

柯比先生表示，除了在中国增长最快的地区之一挑战酿造葡萄酒之外，与国际酿酒师们建立关系是他决定来到宁夏的另一个原因。

他说丽景向中国出口葡萄酒已有8年，并建立了强大的分销网络，在这段时间中，他们已经很清楚强大的关系有多么重要。

"我相信，讲述丽景的故事，以及霍克湾和新西兰葡萄酒的故事，将会成为我在那里工作的一部分。"

宁夏地区出产的葡萄酒在中国和世界都获得了高度赞扬，中国的葡萄酒市场正在蓬勃发展。

醇鉴
宁夏葡萄酒摘金

原文链接：
decanter.com/wine-news/dawa-2015-ningxia-chinese-wines-scoop-gold-medals-273299/

作者：克里斯·莫瑟
2015年9月3日

　　截至目前，宁夏赤霞珠已在2015年亚洲葡萄酒品鉴周上获得两枚金牌，彰显了这个中国葡萄酒产区日益提升的地位。
　　10年前在国外，几乎没有人听说过宁夏葡萄酒，更不用说宁夏赤霞珠了，但在葡萄园和酿酒方面的投资意味着这个中国产区已经引领了新一轮优质葡萄酒生产的浪潮。（节选）

Wine-Searcher
破解中国葡萄酒之谜

原文链接：

wine-searcher.com/m/2015/02/solving-the-chinese-wine-puzzle

作者：克莱尔·亚当森

2015 年 2 月 18 日

  宁夏是最好的地点之一。该自治区的葡萄园位于贺兰山的荫蔽下，沿着黄河排开，那里的气候极具大陆性，但已被驯服利用。高海拔和沙质的冲积土非常重要，而中国强大的人口力量可能是这里风土最重要的部分。银川是一个名副其实的小城市，人口仅有 200 万人，它提供了充足的劳动力以确保葡萄藤在寒冷的大陆性气候的冬季里处于掩埋状态——每株葡萄藤都在秋天被埋土以保温，然后在春天被挖出展藤。（节选）

华尔街日报
法国让位，中国来了

原文链接：
wsj.com/articles/BL-CJB-26768

作者：顾伟（音）
2015年5月7日

随着质量不断的提高，中国葡萄酒已开始在国外赢得奖项。三位葡萄酒专家在《华尔街日报》订户活动的一场盲品会上说，一款来自银色高地酒庄的波尔多风格干红令他们感到惊喜，这家酒庄位于宁夏，宁夏是中国最受人尊敬的葡萄酒产区之一。（节选）

美食家
马蒂亚斯·瑞格在中国出名

原文链接：

nikos-weinwelten.de/beitrag/mathias_regner_big_in_china_ningxia_winemakers_challenge_2015/

作者：凯瑟琳·哈斯
2015年11月8日

你好，干杯！一名年轻的奥地利人参加了2015年宁夏酿酒师挑战赛。

马蒂亚斯·瑞格现年22岁，是"2015宁夏酿酒师挑战赛"中最年轻的参赛者，也是唯一说德语的参赛者。参赛的大多数选手都在30～50岁之间。在中国宁夏的首府银川市，他刚刚收获并加工了他的赤霞珠葡萄。这项比赛的奖金总额为84万元（约合11.5万欧元）！（节选）

印度葡萄酒学院
两名印度人参加宁夏葡萄酒挑战赛

原文链接：

indianwineacademy.com/item_3_675.aspx

作者：萨巴哈·阿罗拉

2015 年 12 月 8 日

中国的宁夏产区正在进行着一项前所未有的葡萄栽培和文化试验，来自 18 个国家的 48 位酿酒师，其中包括印度的 2 位酿酒师——Parikshit Pramod Teldhune 和 Priyanka Kulkarni。他们在为期两年现金奖励超过 11 万美元的酿酒比赛中展示自己的技艺。（节选）

2016

CBS 新闻
中国把荒漠变成葡萄酒产区的豪赌

原文链接：

cbsnews.com/news/china-aims-become-top-wine-producer-ningxia-region-vineyards/

作者：塞思·唐斯
2016年1月1日

"我去过世界上所有的葡萄酒产区，我想，在戈壁沙漠附近生产葡萄酒，不可能吧？不可想象。但是，天啊，戈壁沙漠附近的葡萄酒——这是现实，而且是一个大大的现实。"葡萄酒专家兼作家凯伦·麦克尼尔（Karen MacNeil）说。

现在，麦克尼尔正在更新她的书《葡萄酒圣经》（The Wine Bible），为杂志撰稿，试图理解这些真正的"新世界"葡萄酒。（节选）

年份的足迹Ⅱ：国际媒体报道宁夏葡萄酒

电讯报
中国葡萄酒的红色曙光

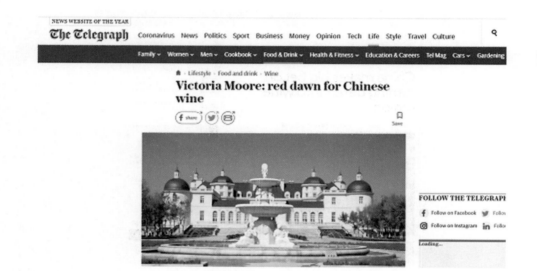

原文链接：

telegraph.co.uk/food-and-drink/wine/victoria-moore-red-dawn-for-chinese-wine/

作者：维多利亚·摩尔
2016年9月17日

　　欢宴集团（Conviviality）的葡萄酒采购总监安德鲁·肖（Andrew Shaw）在春季前往中国时签约了张裕摩塞尔，他表示还打算从其他酒庄购买葡萄酒……

　　"品尝后，我觉得中国葡萄酒的质量可能在两年内翻了一番，而且改进的速度还在加快。酿酒星球上已经到达或将要到达的地方都有一个巨大的潜在盲区，任何率先进入的人都将拥有它。不去那里是我们所不能承受的。"（节选）

醇鉴
中国开展太空葡萄试验

原文链接：

decanter.com/wine-news/china-grows-wine-space-beat-harsh-climate-331421/

作者：吴嘉溦

2016 年 9 月 20 日

中国已经将葡萄藤送入新建并已经在轨运行的"太空宫殿"实验室"天宫二号"之中，以试验葡萄藤的抗旱和抗寒能力……

据宁夏当地媒体报道，这些葡萄来自贺兰山东部地区的一个苗圃，贺兰山东部地区是中国最著名的优质葡萄酒产区之一。

该苗圃为成功集团所有，成功集团自 2013 年开始从法国梅西集团引进葡萄藤。（节选）

世界报
中国宁夏的红色黄金

原文链接：

lemonde.fr/vins/article/2016/10/24/en-chine-l-or-rouge-du-ningxia_5018983_3527806.html

作者：西蒙·莱普雷特
2016 年 10 月 24 日

近年来，当一种中国葡萄酒赢得比赛时，它通常来自宁夏。宁夏位于中国西北部，北邻内蒙古，南邻甘肃。它是一个小自治区，似乎专门生产优质葡萄酒。（节选）

赫曼努斯时报
南非酿酒师在宁夏大展技艺

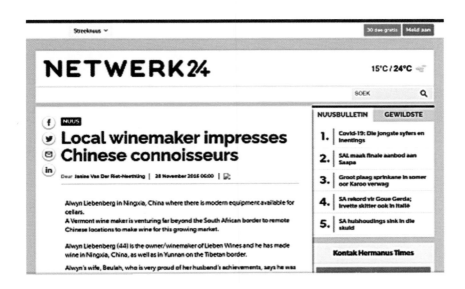

原文链接：

netwerk24.com/ZA/Hermanus-Times/Nuus/local-winemaker-impresses-chinese-connoisseurs-2016 1123-2

作者：珍妮·范·德·里特尼斯林

2016年11月28日

一位来自西开普省佛蒙特的酿酒师跨越南非边境，到万里之遥的中国偏远地区冒险，为这个蓬勃发展的市场酿造葡萄酒。

阿尔温·利本贝格（Alwyn Liebenberg），44岁，是利本（Lieben）酒庄的庄主和酿酒师，他曾在中国宁夏和与西藏接壤的云南地区酿造葡萄酒。

阿尔温的妻子比拉（Beulah）对丈夫的成就感到非常自豪，她说，他去年入围在中国举行的宁夏酿酒师挑战赛的最终名单，为这场为期两年的比赛酿造葡萄酒，这些葡萄酒作品将在2017年9月由一个国际小组进行评判……

比拉说，阿尔温对他酿造的葡萄酒非常满意，因为这款酒具有他想要的波尔多特色。"冬天的宁夏天气如此冷，以至于整个葡萄园都需要被埋进沙土里面，这样葡萄藤就不会被冻死，等天气暖和些后再挖出来。他们甚至在酒窖中也有暖气，以防止葡萄酒在冬天结冰。"（节选）

2017

马尔堡快报
市长访问中国签订姐妹城市协议

原文链接：
i.stuff.co.nz/business/96114564/mayor-takes-trip-to-china-to-sign-sistercity-agreement

作者：詹妮弗·埃德尔
2017年8月30日

新西兰最大的葡萄酒产区即将与中国葡萄酒产区宁夏签订姐妹城市协议。

马尔堡市市长约翰·莱格特（John Leggett）和马尔堡市姐妹城市委员会的代表将在下周的访问期间完成这项协议。

马尔堡市市长约翰·莱格特说，他认为现在的宁夏和30年前的马尔堡有相似之处，当时葡萄在马尔堡还是一种相对较新的作物。

"他们意识到马尔堡现在是世界领先的葡萄酒产区之一，这里蕴藏着大量的知识和专业技能。"

在过去 18 个月里，两个地区进行了多次互访，姐妹城市协议旨在为两个地区创造教育和葡萄酒行业的机会。

到目前为止，已经有一名宁夏酿酒师在马尔堡完成了一个榨季，一群宁夏学生参观了马尔堡女子学院和马尔堡男子学院，还有六名学生进入尼尔森马尔堡理工学院攻读葡萄栽培和酿酒学位。

一些马尔堡葡萄酒技术公司已经与中国客户签订了合同。

马尔堡前市长阿利斯泰尔·索曼（Alistair Sowman）去年与宁夏签署了一份协议。

莱格特说，下一步是通过一份地区对地区的协议使双边关系正式化。

他说："我们已经完成了初步步骤，现在我们理解彼此希望从这种关系中获得什么。"

他说，姐妹地区关系是扩大马尔堡中学和高等教育学院教育与培训服务的大好机会。

委员会成员阿利斯泰尔·索曼（Alistair Sowman）、莉莉·斯图尔特（Lily Stuart）和凯蒂·贝尔（Cathie Bell）将陪莱格特一起出行，时间将从本周日持续到 9 月 10 日。

加入他们的还有酿酒师理查德·奥唐纳（Richard O'Donnell）和戴夫·泰尼（Dave Tyney），他们每年会花一定的时间在宁夏做酿酒顾问，此外还有定居在宁夏的教育代理陈舒。（节选）

买手杂志
中国葡萄酒质量的攀登之路

原文链接:

the-buyer.net/insight/ningxia-china-looks-to-present-quality-wines-to-the-world/

作者:维克多·斯马特
2017年10月2日

没人能说中国人缺乏雄心壮志。因为他们已经让全世界都了解到他们对葡萄酒的重视,中国人已经启动了一个在自己的国家建设葡萄园的大项目,重新定位中产阶级的饮酒习惯,建立管控体系,他们希望能确保他们生产的葡萄酒不仅是一流的,而且还能得到全世界对此举的普遍赞誉。维克托·斯马特(Victor Smart)来到宁夏,亲眼目睹了中国人的雄心壮志。

中国的葡萄酒产业早已成为传奇,宁夏贺兰神国际葡萄酒庄将遍地碎石的荒地变成了3万公顷的葡萄园……哦,同时还建了一个葡萄酒主题公园。

从很多方面来看,宁夏这个被一条狭窄的山脉与戈壁沙漠隔开的经济贫困地区是一个不起眼的地方。凛冽的寒风迫使葡萄种植者要在冬天对葡萄进行埋土防寒,这意味着每隔20年就要把葡萄连根拔起进行植株更新。尽管如此,宁夏是中国生产葡萄酒的新兴产区,对卓越的追求是毋容置疑的。

顶级酒庄在他们的葡萄酒能在质量上与受人尊敬的法国葡萄酒相媲美之前不会停下前进的步伐。一些酒庄的葡萄酒已经能够做到这一点。

宁夏给游客留下的第一印象是有雄心、时不我待和大规模发展。基于当地酿酒师已经酿造出许多优质的葡萄酒和一些顶级的葡萄酒,政府官员们明确表示下一个目标是"让宁夏葡萄酒走向世界"。当地政府投入重金展开宣传,其中包括与澳大利亚联合制作具有故事情节的电视剧……是的,类似与酿酒师恋爱约会这样的情节!此外政府整体希望转型为消费经济,正决心把中国的中产阶级变成一个葡萄酒饮用群体。

未来10年,宁夏葡萄酒的发展对世界其他地区的葡萄酒生产商意味着什么,我们只能猜想。但现在,来自世界各地的酿酒师们聚集在一年一度于宁夏首府银川举行的宁夏葡萄酒博览会上。有瑞典人和英国人,中方的翻译人员也急匆匆地在进行法语学习。正如奥地利酿酒师伦茨·摩塞尔(Lenz Moser)所说的那样"大国屹立不倒"。

在墙面优雅的志辉源石酒庄,我们认识了25岁的庄主袁媛女士,以及她的酿酒师——曾在法国叙兹拉鲁斯学习酿酒的31岁的杨伟明先生。他们两个人管理着130公顷的土地,是中国葡萄酒的新起之秀。

管理葡萄是一项高度劳动密集型工作,而对葡萄进行埋土和出土的工作更是如此。但宁夏在这方面展示出一种极强的天赋和自信。

这两位年轻人生产的霞多丽葡萄酒非常迷人,入口有点肥硕,这让和我一起旅行的法国葡萄酒专家们展开了笑颜。酒庄装备齐全,摆放着中国风的古董,大厅承担着其作为旅游景点的第二功能。

要了解中国的雄心壮志究竟有多大,有必要参观再往北一点的贺兰神国际葡萄酒庄(Ho-Lan Soul)。10年前,该公司的董事长陈德启先生不顾众人的反对,在宁夏贺兰山附近买下了数千公顷布满巨石的荒地。

目前这片葡萄园的栽培面积达到了惊人的3万公顷,是中国最大的酒庄之一。我不知道这里的"风土"应如何概括，但贺兰神的标识中包含了一个令人生畏的传统武士，代表着该品牌征服者的雄心。

洞穴般的仓库里存放着成百上千的装满了不同年份葡萄酒的新法国橡木桶；酒庄的 2014 有机西拉在我们的盲品中表现得特别好，它表现出经典的法国风格，西拉的典型性突出。还有一款上等的 2012 赤霞珠。酒价很高，那款赤霞珠一瓶要花费约 200 英镑。很明显，中国人想要"实打实的昂贵"。

不用说，陈先生的雄心还不止于此。

项目的下一阶段是将现在还很荒凉的土地变成新的葡萄酒旅游路线上的一个吸引人的景点，这将为宁夏回族自治区带来急需的旅游收入。这家酒庄将很快发展成为一个葡萄酒主题公园，私人城堡点缀其间，那些注重品牌的超级富豪有能力购买，甚至有机会种植属于他们自己的一片葡萄藤。

中国已经证明，对于外国酿酒师、装瓶机械制造商和软木塞供应商（对于一瓶昂贵的葡萄酒而言，螺旋盖带来的声誉太低，根本不值得考虑）来说，这里是一个利润丰厚的市场。

一位参加葡萄酒博览会的政府官员高兴地预言，随着中产阶级的壮大，中国将很快成为世界上最大的葡萄酒消费国。美国的葡萄酒消费目前排名世界第一，而中国的人口是美国的 4 倍多，所以实现这个目标尽管说不上容易，但并非不可能。这需要中国的中产阶级从根本上重新定位他们的饮酒习惯，即从传统产品转向葡萄酒，而这一过程已经在顺利进行中。

对酿酒师来说，好消息是名酒的溢价很高。当然，可能最好不要谈论饮酒者的"单位"。

宁夏开始着手葡萄酒列级，效仿波尔多产区闻名世界的可追溯到 1855 年的列级系统，将葡萄酒从第一级到第五级（crus）进行排名。这个举动相比其他更能说明中国与世界上最受欢迎的葡萄酒进行竞争的决心和急迫感。

在欧洲人看来，一个年龄十年多一点的产区要用这样的分级方式来美化自己最好的葡萄酒似乎有些为时过早，这样一个招人嫉妒的目标带来相当大的挑战。但人们怀疑，在中国人眼中，葡萄酒世界不仅只是生产最好的葡萄酒，还要被全球公认为生产最好的葡萄酒。

醇鉴
生物动力法在宁夏

原文链接：

decanterchina.com/en/columns/anson-on-thursday/going-biodynamic-ningxia

作者：简·安森
译者：吴嘉溦
2017 年 10 月 26 日

简·安森（Jane Anson）在本月的一次访问中发现了中国新兴精品葡萄酒产区的另一面。

这是我们宁夏行的最后一天，我们只剩下半小时就要奔往银川河东国际机场。这家酒庄并没有显赫的名声，也并不在我们的访问时间表上，但它正好位于离开宁夏的主干道附近，所以我们决定去看一看。

宁夏回族自治区位于中国人口稀疏的西北，距离北京1 100千米，近年来被认为是发展最快、也最激动人心的葡萄酒产区。

在宁夏的一周，我们访问了诸多风采各异的酒庄。其中有国企旗下的庞大产业——比如中粮集团旗下的长城云漠酒庄，也有国际集团建立的旗舰产业——比如路易·威登－酩悦·轩尼诗集团（LVMH）的夏桐酒庄，还有保乐力加（Pernod Ricard）的贺兰山。此外，我们还访问了著名的小规模精品葡萄酒庄，包括银色高地、贺兰晴雪以及迦南美地。

这些著名的酒庄在贺兰山东麓，海拔在1 200米左右，多位于那条蜿蜒的"葡萄酒文化长廊"之上。

但是，这家酒庄——"博纳佰馥"，却和它们的画风都不大一样。

首先，它非常之小，种植面积仅有2.8公顷，园区面积为6公顷。一对年轻的"80后"夫妇——彭帅和孙淼——是酒庄的主人。夫妇俩在山东上高中时就相识了，毕业于当地的大学后，两人以优异的成绩被选拔参与国际交流项目，飞往法国。

在那里，他们爱上了葡萄酒。两人曾在勃艮第伯恩地区（Beaune，也译作"博纳"）学习，在金丘与生物动力法酿酒师Emmanuel Giboulet并肩工作（译注：这位酿酒师曾经因为拒绝使用杀虫剂抵制虫害而遭到起诉），教皇新堡产区的幽居酒庄（Domaine de la Solitude）也留下了他们的身影。

六年的时光过去，2013年，夫妇俩一同回到孙淼的老家银川，开启了自己的酿酒事业。

"最开始，我们获批的土地位于贺兰山脚附近，但是那里没有电，我们也没有资金建设基础设施。"彭帅说道，"我们转而选择了这个地块。这儿离城市很近，对我们更实际一些。此外，我们还看中了葡萄园周围的树林，这样的环境更有助于促进生物多样性，有利于我们实践生物动力法。"

在宁夏，不乏巨大资本投入建造的酿酒巨擘。例如，保乐力加的宁夏葡萄园，最老的葡萄藤栽种于20世纪90年代。几百公顷沙丘被一点点清理推平，重建耕地，修葺沟渠，才栽上了葡萄园。

宁夏也不乏卢瓦河谷和波尔多式的华美酒庄城堡。在美的控股耗资2 200万英镑建造的美贺庄园，优美的大理石城堡去年刚刚建成开放；张裕摩塞尔十五世酒庄华丽的欧式城堡耗资7 000万英镑，其中包含存储了超过

800个橡木桶的酒窖，还有一个中国葡萄酒历史博物馆。

博纳佰馥似乎走的是截然相反的道路。酿酒厂极小，里面只有四个小型不锈钢发酵罐，和一个敞开的大木桶，用于将花朵、草叶等转变为生物动力法所需的自然肥料。

穿过葡萄园间的泥土路，尽头是只有一个房间的办公室。葡萄园中有几处断垄，孙淼解释道，在寒冬和疾病的侵袭下，新植株每年的成活率只有30%左右。

"我们买的本地葡萄苗，价格只有法国进口苗木的一个零头。"孙淼对我说道。比起英文，她讲法语更加轻车熟路。"现在我们主要想集中精力恢复土壤的活力。我们并不想用大笔的投资和市场推广费用，把自己逼到非成功不可的路上。"

在宁夏的这几天，我很少听到人们这么认真地讨论土壤的结构。

在这里，葡萄园管理依然处于初级阶段，幸而一切正在进步。从几年前开始，张裕摩塞尔十五世酒庄收购葡萄时更重视成熟度，而不是单纯的重量，因为广受尊敬的奥地利酿酒师伦茨·摩塞尔给张裕葡萄园带来了更专业的管理方式。

美贺酒庄则邀请葡萄种植专家Richard Smart担任顾问，指导酒庄更好地管理葡萄园。近年新建成的酒庄多采取"厂"字形修剪方式，以确保葡萄能够更均匀地成熟，倾斜的主干则令冬季埋土造成的损伤降到最低。不过，多数酒庄还是致力于将大笔费用投注在闪亮的现代化酒窖中。

在博纳佰馥，最老的葡萄藤种植于2010年，是两夫妇在法国工作期间，趁回国的间隙种上的。

比起贺兰山脚贫瘠得惊人的碎沙石土壤，这里的土壤黏土含量更高，但土壤中有机物含量依然极低，pH则高达9。

Nicolas Joly的女儿Virginie曾两次为酒庄提供生物动力法所需的生物制剂BD500，其他必须的制剂则来自马孔（Macon）的一家生物动力制剂的供应商。

孙淼说，他们最近在宁夏找到了一家有机养牛场，这样牛角[*]就有了着落。就在我们走过葡萄园时，来自这家养牛场的新鲜牛粪已经送到，正

---

[*] 牛角中填满牛粪，在秋季埋入土壤中，第二年春天挖出，播撒在葡萄园中。

被播撒在葡萄园里。

夫妇俩兴致勃勃地谈论着土壤和微生物，似乎对"地下"发生的一切都充满兴趣。毫无意外地，博纳佰馥唯一可谓"大手笔"的投资，便是地下酒窖了。"这个酒窖，我们只花了一个月就建好了，这就是中式效率！"彭帅说道。

酒窖的入口，看起来像水泥建成的地下掩体。钻过门帘，潮湿的泥土、蘑菇的气味和清凉的空气迎面扑来，令人恍如置身法国的酒窖之中。眼睛适应了昏暗的光线之后，可见两行橡木桶一路延伸，尽头是整齐码放的酒瓶，瓶身都标注着年份和风格。长品酒桌的四周，是用酒桶制成的椅子。

在这里，我才第一次看到了酒庄的英文名："Domaine des Arômes"。中文名"博纳佰馥"则由孙淼的父亲亲笔书写。地下酒窖没有控温设备，尽管室外冬季气温可低至 -25℃，夏季气温可高达 40℃，酒窖里却全年保持 8～18℃。

在本次旅程中，我品鉴了许多款宁夏葡萄酒，其中不乏令人印象深刻的美酒。但令我单纯为发掘它们而感到快乐的，却为数不多。

我们只品鉴了两款 2013 年份的红葡萄酒，以及 2014 年份的红葡萄酒和白葡萄酒各一款。其中"馥"是自然酒，其他三款采用正常工艺。"博纳佰馥"是赤霞珠与美乐的混酿，正如宁夏产区众多的红葡萄酒一样。这些酒还在发展途中，需要时间，等葡萄园的土壤焕发更多生命力的时候，才能长足进步，彰显风骨。但是我感受到了这些酒中的匠人之心，以及令人期待的未来。

年份的足迹 II：国际媒体报道宁夏葡萄酒

北京人
品鉴中国葡萄酒的完美入门清单

原文链接：

thebeijinger.com/blog/2017/02/15/annual-grape-wall-challenge-gives-perfect-great-hit-list-chinese-wines

作者：吉姆·博伊斯

2017 年 2 月 15 日

　　从事食品行业市场营销的崔云安（音）说，本次品鉴展示了每一款葡萄酒的独特性……

　　她对家乡宁夏的葡萄酒感到惊讶。"我知道我们那里的气候适合葡萄生长，我也知道我们能酿造出优质的葡萄酒，但我不知道我们可以酿造出如此惊艳的葡萄酒！"她说，"这超出了我的期待。"（节选）

中国日报亚洲版
各国酿酒师参加宁夏葡萄酒挑战赛

原文链接：

chinadailyasia.com/asia-weekly/article-13555.html

作者：王浩（音）、李翔（音）

2017年10月16日

  他们从全球各地来到这里，他们所具有的知识、智慧和耐心将帮助他们及其东道主应对摆在他们面前的挑战。对于来自16个国家的48位酿酒师而言，位于中国西北部的遥远的宁夏回族自治区不仅成了他们的第二故乡，而且在过去的两年中是他们酿造完美葡萄酒的测试实验室。

  ……最大的赢家是以贺兰山东麓为核心的宁夏葡萄酒产业，宁夏的酿酒师在过去两年中，从实际工作中获得了必不可少的专业知识和技巧。

  （节选）

2018

WineLand
超越纳帕，宁夏驾到

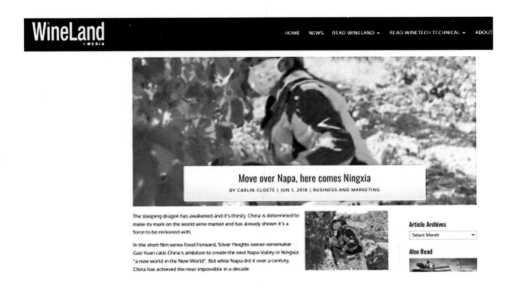

原文链接：
wineland.co.za/move-over-napa-here-comes-ningxia/

作者：卡林·克洛特
2018年6月1日

沉睡的巨龙醒来了，渴了。中国决心在世界葡萄酒市场上留下自己的印记，并且已经展示了它是一支不容忽视的力量。

在《食想》（Food Forward）系列短片中，银色高地的庄主兼酿酒师高源称中国雄心勃勃地在宁夏创建下一个纳帕谷"新世界中的新世界"。纳帕用了一个多世纪才实现目标，而中国在10年内就实现了这个近乎不可能的壮举。

2015年，路透社报道称中国已超过法国，成为世界第二大葡萄种植区（包括鲜食葡萄）。这是一个飞速的增长！根据国际葡萄与葡萄酒组织

（OIV）的数据，自20世纪初以来，中国的葡萄种植面积增加了一倍，达到近80万公顷。2015年，中国占世界葡萄种植总面积的10.6%，而法国为10.5%，世界排名第一的葡萄种植国西班牙为13.5%。

"这一切都是由政府推动的。"利本葡萄酒公司（Lieben Wines）的阿尔温·利本贝格（Alwyn Liebenberg）说。他在中国从事咨询工作，担任飞行酿酒师，将于今年晚些时候第九次返回中国。"例如，政府已经决定，曾经是一个相当贫困的煤矿区的宁夏，将成为世界的下一个波尔多。宁夏毗邻戈壁沙漠，由于中国大多数人口居住在沿海地区，他们希望使这一地区具有发展的吸引力，以便国家的整体发展。为了推动葡萄酒生产商的起步，政府经常向他们提供软性贷款，其中只有一部分需要偿还。所有的大公司都在这里：路易·威登-酩悦·轩尼诗、保乐力加和许多澳大利亚的大品牌。"

宁夏的土壤是葡萄栽培的理想之地，尽管处于干旱的半沙漠地带，但黄河流经该地区。在一些地方，要钻200米深才能找到充足的淡水。在政府的帮助下，人们修建了水坝，安装了复杂的灌溉系统，将荒芜的土地改造成连绵不断的葡萄园。根据2015年发表的一篇《纽约时报》的文章，政府计划到2020年在这里建成200家酒庄。

像波尔多一样，宁夏位于北纬38°。高源，也被称为艾玛（Emma），在法国学习了酿酒技术后，她与法国丈夫吉利（Thierry Courtade）回到家庭农场，负责经营银色高地酒庄。

在《华尔街日报》的一次包括有首位亚洲葡萄酒大师李志延等多名经验丰富的品鉴者参加的盲品中，银色高地的赤霞珠葡萄酒和来自山西怡园酒庄的赤霞珠葡萄酒的表现都超过了教皇新堡等知名葡萄酒，这令品酒者大吃一惊。如果葡萄园管理得当，戈壁沙漠干燥、沙砾覆盖的土壤似乎可以生产出伟大的葡萄酒。

"有一些怡人的葡萄酒产自这个地区，"阿尔温说，"但我认为中国还有更好的葡萄酒产地。你不会像在南非那样，在60兰特一瓶的酒中找到这种品质。但花到3000兰特，你真能见到一些超棒的葡萄酒。"他最近为一家法中合资的公司做咨询，在中国的藏区酿造葡萄酒。"我住在山里，周围住着只会说藏语的妇女。他们有古老的酿酒文化，从采摘到分拣，一切都是手工操作，葡萄酒都装在瓮里。这些酒的价格（便宜得）太离谱了，所有产品都在上海市区被抢购一空。"

当然,如果没有市场的协同发展,单种植优质葡萄是没有用的。努力使自己成为下一个纳帕,意味着要说服国际葡萄酒买家,让他们感兴趣。而中国人正以创新的方式来实现这一目标。山脊酒庄(Mountain Ridge)的贾斯汀·科兰斯(Justin Corrans)在2015年应邀参加宁夏国际酿酒师挑战赛时,对中国一无所知。这是一个出色的营销策略,使宁夏成了一个广受瞩目的顶级酿酒地区。贾斯汀、阿尔温和同样来自南非的卡斯滕·米利亚里娜(Carsten Migliarina)是来自17个国家的48位酿酒师中的三位,他们被邀请参加为期两年的酿酒挑战赛。每位参赛者都与当地的一家酒庄配对,挑战内容就是酿造葡萄酒。贾斯汀与兰轩酒庄配对,去年8月29日,一个由国际评委组成的葡萄酒评审团宣布贾斯汀以他酿造的赤霞珠葡萄酒位列五个金奖之一并获得了挑战赛的总冠军。

"在沙漠中酿酒的挑战之一是难以置信的温度波动。"贾斯汀说,"冬天常常会冷到-20℃,他们不得不埋下葡萄藤来保护它们。"去年,中国将赤霞珠、美乐和黑比诺葡萄藤送入在太空的"天宫二号"实验室,希望引发突变,使它们更适合恶劣的气候并具有抗旱和抗病毒的能力。

气温波动不是宁夏唯一的问题。"中国现在正面临着南非在30年前所面临的同样问题,"贾斯汀说,"他们正在努力获得优良的品系。请记住,他们都是第一代的葡萄酒饮用者、酿酒师和葡萄栽培师,因此他们必须广泛咨询以弥补经验的不足。他们适应得非常快,而且这个行业实际上只有10年的历史,所以要获得好的无病毒植株还是个问题。"

《葡萄酒经济学家》(The Wine Economist)博客的编辑迈克·维塞思(Mike Veseth)写道,在中国要掌控土地并非易事。许多葡萄酒生产商仍不得不与数百个家庭经营的小葡萄园合作。一些公司正通过与葡萄生产者建立关系来应对这一困难的供应链。他们将土地租给自己的员工,提供培训和信贷以换取投入。

法国的英语频道24电视台(France24 English)的一份报告显示,自2013年以来,中国已经成为全世界最大的红葡萄酒消费国。斯皮罗斯·马兰达克斯(Spiros Malandrakis)在发布于Just-Drinks网站的2018年葡萄酒行业主要趋势的摘要中预测,中国将加速从把葡萄酒作为奢侈品向大众市场销售的转变，"这种发展将进而从供应与需求、品牌与市场定位两个方面塑造全球葡萄酒产业"。

贾斯汀和阿尔温都认为中国人是真的喜欢法国酒。贾斯汀·科兰斯说："如果听起来是法国的，那一定很棒。"因此很重视诸如赤霞珠这样的法国经典品种。但贾斯汀补充说，他认为，在宁夏如果他们用早熟一些的品种如美乐或黑比诺，酿出的葡萄酒能好很多。

《醇鉴中国》的定期撰稿人李德美说，马瑟兰是另一个潜力巨大的品种。李德美曾是中法庄园的酿酒师，他在法国顾问的推荐下酿出了中国的第一款马瑟兰葡萄酒。"我相信马瑟兰的潜力在于其高产量、对环境的适应能力、复杂而美丽的风味、丰富而集中的口感，以及酿造不同风格葡萄酒的潜力。"

他在一篇关于中国葡萄酒种植气候的文章中说，在中国通常被称为"葡萄酒产区"的地区实际上是行政区划。葡萄酒酿造历史最长的是山东省，尤其是山东半岛及其下辖烟台地区。从产量和产值来看，山东省和河北省的葡萄酒占中国葡萄酒产业的一半以上。其他酿酒地区还有北京、天津、山西（另一家国际知名酒庄怡园就坐落在那里）、陕西、吉林、辽宁、新疆维吾尔自治区、宁夏回族自治区、甘肃、内蒙古自治区以及云南和四川省，后两者与西藏接壤。

如此惊人的增长曲线能否持续，还有待观察。令人欣慰的是中国公众正越来越多地将葡萄酒作为日常饮料，从而扩大了购买葡萄酒的人群，这只会对全球葡萄酒产业有利。

中国日报
铁路上的葡萄酒

原文链接：
chinadaily.com.cn/a/201808/17/WS5b760ecaa310add14f3863c5_1.html

作者：李映雪（音）
2018年8月17日

得益于新推出的豪华列车服务，葡萄酒爱好者们现在可以踏上宁夏葡萄园的探索之旅，并一路品尝宁夏的美酒。

全体上车！全体上车！晚上8点开往银川的列车有一个不容错过的特别之处——加挂了一节额外的精品车厢，里面有一个专门定做的品酒室和六个软卧包间。

在这趟列车上的一晚有些与众不同。在这节豪华车厢里，旅客们可以享用一顿搭配贺兰山东麓葡萄酒产区美酒的晚餐，睡前还可以敷上葡萄酒面膜。

本项目于7月31日开通,这是中国第一家移动酒庄,也是北京与宁夏回族自治区首府银川的首个夜间豪华铁路班列。

车厢内有两间单人套间、两间带共用盥洗室的双人卧室和两间四人房,可容纳14名旅客,所有房间都免费提供洗漱用品和毛巾。

移动酒庄项目是宁夏葡萄产业发展局的创意,由中国铁路兰州局集团运营。

来自全国各地的200名记者受邀在8月份体验这个项目。根据项目协调员杨洋(音)的说法,该项目将在未来三年作为一项客运服务。

"我们给这节车厢配备了一个单独的厨房,我们会邀请贺兰山东麓葡萄酒产区不同酒庄的厨师来轮流为乘客做饭。"杨洋说。

侍酒师将葡萄酒与菜肴搭配,并将他们的葡萄酒知识传授给旅客。

"火车于次日早晨到达银川,乘客们可以参观当地的葡萄园,采摘葡萄,品尝不同的葡萄酒。"杨洋说。

贺兰山东麓葡萄酒产区于2002年被国家质量监督检验检疫总局(现国家市场监督管理总局)认定为国家地理标志保护产区。

2013年,该地区推出了其首个葡萄酒列级制度,以提高对中国优质葡萄酒生产的监管。

该地区目前有86家酒庄,葡萄种植面积超过3.8万公顷,年产葡萄酒约10万吨。这些酒庄创造了大约12万个就业机会,每年吸引40多万的游客。

立兰酒庄的酿酒师邵青松表示宁夏是种植葡萄的理想地点。"我们宁夏雨水不多,如果需要用水,我们可以从黄河调水。"邵青松说。

立兰酒庄位于银川西南30千米的闽宁镇原隆村,地处贺兰山东麓葡萄酒产区的中心地带。

黄河水富含有机质,有利于葡萄生长,土壤在夜间释放热量,有助于葡萄果实成熟。

"我们在葡萄上使用自己生产的有机肥,而不是化肥,这有助于降低成本,同时提高质量。"邵青松说。

与其他葡萄酒产区不同,冬天宁夏的葡萄藤必须被埋在土里,第二年春天再展藤,以防止它们在极端干冷条件下死掉。邵青松和他的同事们正在努力寻找解决这一问题的方法。

"在107公顷的葡萄种植面积上,我们可以生产大约500吨葡萄,可以酿制成大约40万瓶葡萄酒,"邵青松说,"我们的葡萄酒现在销往法国和英国。"

邵青松认为宁夏有能力生产出高水平的葡萄酒,但还需要时间来提高。"如果你今年种下葡萄,5年之后产出的葡萄才能有足够的品质,确保酿造出高质量的葡萄酒。我们的酒庄才酿造了3年葡萄酒,所以还有很长的路要走。"

成立于1997年的贺东庄园没有遇到同样的问题。他们的葡萄园有一些成年的葡萄藤,其中一些可以追溯到一个多世纪以前。

在2010年接手酒庄后,酒庄庄主龚杰前往清华大学参加MBA课程,进一步学习如何管理自己的葡萄园。

"我过去是做采矿生意的,所以我必须从头开始学习酿酒,"龚杰说,"自从我开始和当地农民一起耕耘土地,从春天的最早的萌芽到秋天的收获,生活一直在稳步改善。

"当地政府也支持我们,在那些最具影响力的葡萄酒大赛上,每获得一个奖项,政府就会给予酿酒师50万元(72 600美元)的奖励。到目前为止,我们已赢得了三个奖项。"

龚杰正在他的酒庄旁边建一个"葡萄酒小镇",他希望这将成为一个AAAA级的旅游景点(AAAAA是国家最高级别)。通过开设葡萄酒文化课程和建造一家舒适的酒店,龚杰希望游客能在他的葡萄园里待得更久,并更多地了解他的葡萄酒。

龚杰说:"我们不得不将参观我们酒庄的游客人数限制在每天200人左右,因为酒窖里的人多了,可能会影响酒窖的温度和葡萄酒的质量。"

大多数位于贺兰山东麓产区的酒庄都是旅游景点。

目前,该地区正与一家游轮公司商讨,像在火车上一样,也在游轮上建造一个酒庄。

宁夏葡萄酒产区在2013年入选《纽约时报》"必去的46个地方"榜单,《纽约时报》的推荐词是"当地政府开垦了荒漠化的土地,种植了赤霞珠和美乐葡萄,慷慨地进行灌溉,并发起了一场运动,要把这个崎岖的落后地区变成中国的波尔多"。

所以现在，只需花上一张火车票的钱，葡萄酒爱好者们就可以踏上一段探索之旅，沿途品尝宁夏葡萄酒之乡的味道，领略张裕摩塞尔十五世酒庄的欧洲风格，或者参观志辉源石酒庄的中国传统沙漠建筑。

美食家
穿越中东的葡萄酒之路——
中国最优葡萄酒

原文链接：

nikos-weinwelten.de/beitrag/auf_der_wein_route_durch_das_reich_der_mitte_beste_weine_chinas/

作者：尼克·罗森博格
2018 年 8 月 19 日

  第一次站在埃菲尔铁塔和金字塔前时，它们看起来并没有想象中的那么大。由于有很多关于它们的明信片、电影和故事，你以为它们会很大，于是你可能会有所失望。你知道这种现象吗？
  中国却不是这样的！尽管同事们的报告中所陈述的规模令人难以置信，斯图尔特·皮戈特（Stuart Pigott）和中国通弗兰克·卡默（Frank Kämmer）

多年来一直在向我转述，现在我终于要穿过中东。令人震惊的事情发生了：在中国，一切真的都变得更大、更不可思议、更令人兴奋！

在约一周的时间里，我们只探索了中国葡萄酒之路的一小部分。宁夏是中国的纳帕，酒庄的规模与迪斯尼家族庄园西尔弗拉多（Silverado）的规模相同，在那里运动型多用途汽车（SUV）可以轻松地停放在狭小的地方。宁夏的许多酒庄看起来好像是直接从法国卢瓦尔河地区进口的，建筑有7层而不是2层，并且还配备了12座塔楼，而不是4座塔楼。

宁夏正在逐渐发展成为中国最好的葡萄酒产区。我们参观了那里的9个酒庄，并品尝了来自"宁夏葡萄酒挑战赛"的数十款葡萄酒。宁夏的葡萄酒之路全长150千米，有100多家酒庄，种植面积约为3万公顷。产区离银川市不远。银川是一个拥有200万居民的城市，在北京以西约1 100千米，内蒙古戈壁沙漠的南边。

这里夏季的气候非常理想。但在冬季，最低温度可达-25℃，葡萄藤的生存变得很艰难。因此，葡萄藤会被倾斜种植，在冬季它们会被掩埋，并覆盖土壤以进行保护。通常，这些地区会浇冻水，葡萄园的土层被冻结，葡萄藤可以在冰冻的土层下生存。

张裕先锋葡萄酒集团由外交官张弼士于1892年创立，他的第一位酿酒师是奥地利人奥古斯特·威廉·巴伦·冯·巴保（August Wilhelm Baron von Babo）。如今，张裕已成为中国最大的葡萄酒酒庄品牌，现在最重要的当地顾问又是一位奥地利人伦茨·摩塞尔（Lenz Moser），他曾在罗伯特·蒙大维（Robert Mondavi）酒庄工作了多年，现在为张裕经营摩塞尔十五世酒庄。

此外，其他酒庄如张裕爱斐堡国际酒庄、张裕卡斯特酒庄、张裕瑞那城堡酒庄和张裕巴保男爵酒庄都位于中国。在烟台附近的张裕国际葡萄酒城，有95个葡萄酒生产区。感觉每个分区都像柏林滕珀尔霍夫机场的前候机厅一样大——总计27万米^2，约有250个足球场那么大。

它是世界上最大的葡萄酒生产商。据推测，在满负荷生产情况下，每年将产出40万吨葡萄酒和烈酒，相当于德国葡萄酒年产量的一半（介于750至950万升之间）。

在中国，大多数葡萄酒（超过90%）为红葡萄酒，赤霞珠占大多数，其次是美乐和西拉。我们能够品尝到的所有葡萄酒都取得了巨大的成功，这些葡萄酒经过了纯净的酿造，其中最好的葡萄酒会在波尔多或罗纳河谷的品尝中引起令人瞩目的轰动。

普通葡萄酒的价格每瓶为 10～20 美元，更好的价格能升到 50～100 美元，甚至 150 美元。对于许多中国富豪和大城市中不断壮大的中产阶级来说，这不是个问题，因为他们也希望有优质葡萄酒和酒吧。

据称，超过 90% 的葡萄酒是在中国本土消费的。但大多数中国人仍然喝啤酒和廉价酒。我对大量葡萄酒售卖的假设是：将来，中国葡萄酒会以较低的价格供应给欧洲，中国葡萄酒将会通过新的丝绸之路和海上丝绸之路被运往世界各地。

德国铁路公司宣布：到 2020 年德国和中国之间的货物运输量将达到 10 万个集装箱。总投资规模达 9 000 亿美元的"一带一路"首批线路已经建成。2018 年 5 月初，首列装载 44 个集装箱的货运列车在中奥直达 9 800 千米的新线路上抵达维也纳，中国葡萄酒将会受到极大的欢迎。

中国的葡萄园面积接近 100 万公顷。这意味着中国的葡萄园面积仅次于西班牙。但是，葡萄园中的葡萄种植也被广泛用于生产廉价酒、鲜食葡萄和葡萄干。尽管如此，在短短几年内，中国将成为世界上最大的葡萄和葡萄酒生产国。10 年前，中国的人均葡萄酒消费量为 0.2 升，现在为 1 升，而这种趋势还在上升。

这对宁夏葡萄酒之路意味着什么？再过 10 年应该有 60 000 公顷的种植面积和 300 个中国酒庄，也就是说翻了一番。还有疑惑？那就去宁夏品尝一下葡萄酒吧！

宁夏最佳酒庄的美食世界评选

重大事件：来自中国的葡萄酒海啸很快将会席卷到我们这里
穿越中东的葡萄酒之路

1. 张裕摩塞尔十五世酒庄——张裕酒业的一部分，是中国最好的酒庄之一。非常物有所值，2015 年的旗舰赤霞珠轻松获得 93 分。

2. 留世酒庄——很有潜力！传统酿酒葡萄品种赤霞珠得到 93 分，这位年轻的留世酒庄酿酒师已经是中国的酿酒天才！

3. 银色高地酒庄——庄主高源和她的丈夫吉利（Thierry Courtade）曾在波尔多凯隆世家（Château Calon-Ségur）工作，银色高地每年生产约50 000瓶酒。霞多丽白葡萄酒非常好，是该地区最好的霞多丽葡萄酒之一，获得92～93分。品丽珠在贺兰山也非常优秀。

4. 加贝兰（贺兰晴雪酒庄）——2009年贺兰晴雪酒庄的张静用赤霞珠酿造的加贝兰珍藏葡萄酒赢得了醇鉴葡萄酒大奖，这成为中国葡萄酒的一个里程碑。加贝兰是非常优秀的葡萄酒！

5. 迦南美地酒庄——充满基督教色彩的酒庄，由酿酒师王方管理和推广。小马驹干红葡萄酒、小野马红葡萄酒和黑骏马红葡萄酒令人信服，品质不断提高，而且真的很有趣！

6. 立兰酒庄览翠红葡萄酒——时装界老板精心酿制的葡萄酒。

7. 东麓缘酒庄——正在崛起，这是该地区的新建酒庄，可以说，它的旗舰产品是质量非常不错的家族珍藏。

8. 志辉源石酒庄——这些葡萄酒被命名为"山之子"和"山之魂"，很好地描述了那些仍需进步才能在欧洲获得成功的葡萄酒。

9. 夏桐酒庄——虽然还没有像我们这样的香槟级别，但他们已经开始酿造起泡酒了。四款酒的价格在每瓶20～25美元。

罗宾逊网站
宁夏的青春锋芒

原文链接：

jancisrobinson.com/articles/young-guns-of-ningxia

作者：露易丝·哈伦

2018年10月2日

露易斯·哈伦（Louise Hurren）通过对宁夏葡萄酒产业五位突出的成功人士的描述，展示了中国葡萄酒的快速进化。

在中国的六个主要葡萄产区中，宁夏回族自治区的特色在于当地政府大力支持葡萄酒生产。

宁夏位于北京西南683英里（1100千米）处，已经拥有约100个酒庄（法国重量级路易·威登-酩悦-轩尼诗集团和保乐力加都在这里投资），葡萄园面积达到约4万公顷（98 850英亩），沿着贺兰山东麓展开。

干旱的沙质土壤上种有未经嫁接的葡萄藤（主要是红葡萄品种，通常为赤霞珠、品丽珠、马瑟兰、美乐和西拉，霞多丽和雷司令是最常见的白葡萄品种）。

自 2012 年启动"宁夏国际酿酒师挑战赛"（一个资助希望在该地区磨练技能的外国酿酒师的项目）以来，宁夏得以收获日渐增加的媒体报道。当地政府正在大力推动葡萄酒产业的发展，他们的目标是到 2020 年拥有 6.6 万公顷的葡萄园，并举办亚洲葡萄酒与烈酒丝绸之路大会、葡萄酒展和葡萄酒比赛。

第二届亚洲葡萄酒与烈酒丝绸之路大会于上月在省会银川举行，来自全亚洲（中国、格鲁吉亚、阿塞拜疆、土耳其、摩尔多瓦、亚美尼亚和以色列）的近 800 款葡萄酒参加了比赛。在中国葡萄酒获得的 108 枚奖牌中（本次大会共评出 264 枚奖牌），有 65 枚由宁夏葡萄酒夺得。中国的其他获奖地区分别是新疆、河北、山东、甘肃、河南、吉林、辽宁、内蒙古、天津和北京。参赛葡萄酒由来自 18 个国家的 49 位葡萄酒品鉴大师［包括莎拉·简·埃文斯（Sarah Jane Evans）、安妮特·斯卡夫（Annette Scarfe）和佩德罗·巴列斯特罗斯·桃乐丝（Pedro Ballesteros Torres）］品评。宁夏葡萄酒的表现相当不错：在 18 个大金奖中，他们获得 11 个，该地区总共获得了 132 枚奖牌中的 65 枚。

许多年轻的、精通葡萄酒的中国人以侍酒师、酿酒师、葡萄酒教育者和营销人员的身份参加了今年的丝绸之路大会，他们用流利的英语介绍自己，使用英文名——尽管对我选择的下面五位最令人印象深刻的年轻人，我用他们的中文名字，按照中国的习惯姓氏在前。这几位年轻人热情洋溢，预示着宁夏葡萄酒的未来。

任艳玲

任艳玲是保乐力加贺兰山酒庄的首席酿酒师兼生产经理，她 18 年前以酒窖员工的身份加入酒庄，见证了酒庄从国有到合资，再到 2012 年由保乐力加完全拥有的过程。如今，她管理着一个 20 人的团队。

41 岁的琳达（她和说英语的人交谈时用的名字）是一位葡萄园经理的

女儿，她在当地长大，但她去过法国、意大利、澳大利亚和新西兰，她在霍克湾（Hawke's Bay）的教堂路经历了2016年的榨季。她计划今年晚些时候前往波尔多参加法国国际葡萄酒及蔬果栽培设备技术展（Vinitech）。

当我见到她时，她正忙着监督采收。贺兰山品牌旗下的132公顷（325英亩）葡萄是人工采收的（杂草控制也依靠人工方式），在葡萄园劳动力供应充足的宁夏这是产区的典型管理方式。同样具有代表性的是，该地区冬季寒冷（温度常为-5ºF/-20ºC），意味着从11月至第二年3月葡萄藤必须被沙土覆盖以保护越冬。

然而，由于劳动力的老龄化和该地区的城市化，以及中国独生子女政策近期的变化，琳达预期未来这里将发生变化。因此，所有新种植的葡萄都改用适合机械操作的架式。她说："我们正在为未来做准备。"

贺兰山品牌制定的宏大目标是在2025年左右将产量提高10倍，因此发展变化非常快。每年都会种植更多的葡萄藤（西拉、马瑟兰和马尔贝克现在都在当地的苗圃中生长，未来它们将加入已经开始生产的霞多丽、美乐和赤霞珠的队伍）。

新的办公室于2017年建成，对灌装线的改造也已完成。在原来容积为32 000升的不锈钢罐中间增加了一些较小尺寸的不锈钢罐。"我们每年都会增加一些新的不锈钢罐，使我们的产能可以不断提升。"琳达解释道。

自2016年以来，她一直专注于葡萄酒的品质和研发，尝试使用氧气和整串压榨技术。她还建立了新的团队，采用了新的管理工具（在酒庄的墙壁上有详细说明任务和目标的图表，一块巨大的写字板展示着更多的建设计划）。

关于风土问题，琳达说："宁夏葡萄酒还没有统一的风格。它既不是法国的旧世界风格，也不是澳大利亚的新世界风格。在这里，一些酒庄采用传统方法，另一些则使用新方法，但都有一些共同点，例如浓郁的颜色和良好的酒体结构。但是仍然有很多工作要做，对我们贺兰山品牌来说，我们正不断提升果香。"

贺兰山品牌在今年的丝绸之路葡萄酒大赛中获得了一个大金奖、三个金奖和一个银奖，这也证实她干得很棒。

陈朱云

朱云（或艾丽卡，这是她自信地用英语介绍自己时使用的名字）代表宁夏迦南美地酒庄出席了丝绸之路葡萄酒大赛。31岁的她面带微笑，她曾在中国学习德语，之后在德国的法尔茨地区学习了四年葡萄栽培和酿酒学，包括在伯格多尔特酒庄两年的工作和在德国国际葡萄酒大赛担任评委的经历。

在德国期间，艾丽卡向当地的葡萄酒爱好者和她的同学展示了中国葡萄酒。"他们对中国葡萄酒的产量和质量都感到非常惊讶，同时也对大多数酒的价格感到震惊。"她笑着说（迦南美地葡萄酒的出厂价从每瓶20欧元到85欧元不等）。

2016年她回到中国，从事葡萄酒推广、教育和葡萄酒活动组织工作，在上海附近的一个小城市经营着自己的公司。和许多年轻的中国人一样，艾丽卡与社交媒体联系密切，并对社交媒体感到自在，相信社交媒体在葡萄酒行业的应用潜力。她说："社交媒体对营销推广尤其重要，对于那些希望建立自己形象并与消费者互动的酒庄来说更是如此。"

在众多可用的应用软件中，她推荐了微信（流行的即时通信软件，具有多种社交功能，包括朋友圈，时间线的感觉与脸书类似）、新浪微博［有许多意见领袖（在中国被称为KOL）入驻和使用的网站］和抖音（于2016年推出的新应用）。

艾丽卡在丝绸之路展会的各个展位上品尝了一圈之后，谈到了宁夏葡萄酒高酒精含量和果味甜感的发展趋势，她将此归因于该地区的大陆性气候，以及橡木的普遍使用。

不过，她也聊到品鉴中的细微差别。她说："对我来说，品尝这些葡萄酒有些挑战，因为我的口感已经习惯了德国葡萄酒，德国酒更新鲜、更优雅，但我仍为我在宁夏的所见所闻感到惊喜。中国是一个年轻的葡萄酒生产国，我为这里酿造的葡萄酒感到骄傲。"

李航

李航，英文名字是史蒂文（Steven），今年31岁，是政府认证的中国侍酒师学院的联合创始人。他在中国学习酒店管理，在迪拜和阿布扎比工作了5年，期间获得了WSET 3级证书，以及侍酒师学院颁发的证书，并在返回中国之前在美国参加了为期6个月的侍酒师课程。

中国侍酒师学院每年培训1 500名侍酒师，组织约450名学生到宁夏进行实地考察。学院还组织编写了有关宁夏葡萄酒的培训内容。史蒂文定期举办美食与美酒搭配晚宴，在宴会上他同时展示进口葡萄酒和中国葡萄酒。他说："宁夏葡萄酒必须被包括在内，以证明中国可以酿造出高质量的葡萄酒。"

他认为，社交媒体是在中国推广葡萄酒的关键。"今年我上传了一个侍酒师学院的葡萄酒视频，在短短的三天内，它获得了1 800万的播放量。"侍酒师学院的招生显然也主要是通过此渠道完成的。

传统中医不鼓励喝冷饮，吃饭时服务员通常倒上温水，瓶装啤酒也经常是在室温下被喝掉的。因此，在培训员工从事接待服务和零售工作时，史蒂文要花时间示范侍酒温度的影响，员工掌握这些知识可以使他们更好地为顾客提供葡萄酒消费建议。

在丝绸之路大会上谈到食物和葡萄酒的搭配时，他提到历史上红葡萄酒一直是中国葡萄酒消费者的首选，但是"由于中餐菜肴风味的复杂性以及我们倾向分享菜肴的方式，白葡萄酒通常是比红色更和谐的选择"。在过去的三年中，他看到了饮酒习惯的转变："葡萄酒教育在不断发展，消费者对葡萄酒的了解越来越多。一线城市的白葡萄酒和起泡酒消费量正在增长，贸易展览会的反馈表明，一些进口商和零售商开始对这些产品更加重视。例如，在北京和上海，现在有专门经营白葡萄酒和起泡酒的公司。我坚信，随着市场的不断发展，在中国这些类别葡萄酒的销售将继续增长。"

同样，他观察到宁夏葡萄酒的生产发生了巨大变化（"质量一直在不断提升，现在有许多酒庄获得了国际奖项"）。他觉得这个年轻的地区具有举世闻名的潜力，尽管"宁夏产区还有很长的路要走，例如要研究风土，并试验新的葡萄品种"。他预测气候变化会带来一些影响，因为很少有酒庄有足够的经验来应对这种变化。此外，他还建议需要在市场营销领域进一步取得进展（"酒庄需要开始专注于品牌建设与推广"）。

要克服的障碍包括成本（"建立酒庄的初期投资高且机械化程度低，这些精品酒庄每年的产量为 5 万～10 万瓶，因此总体成本相对较高。与进口葡萄酒相比较，我们的价格没有竞争力"）和消费者对宁夏葡萄酒的认知（"一些中国人仍然认为无论价格和质量，外国葡萄酒比国产葡萄酒好"）。

刘爱国

路易·威登－酩悦·轩尼诗（LVMH）的夏桐酒庄距离银川市中心仅 40 分钟车程。自从 2012 年宁夏夏桐项目启动以来，刘爱国（艾伦）就一直在这里工作。宁夏夏桐与 LVMH 在澳大利亚、巴西、美国、印度和阿根廷的起泡酒生产项目一样，是 LVMH 全球产品的一部分。

之前，他曾在内蒙古的汉森酒庄工作，担任酿酒师兼技术经理。他于 31 岁时加入夏桐酒庄，担任运营酿酒师，负责酒庄种植的 68 公顷（168 英亩）黑比诺和霞多丽，每年生产 70 万瓶起泡酒。据他介绍，这是在中国首次实现量产的传统法的起泡葡萄酒。

艾伦拥有中国西北农林科技大学的葡萄栽培和酿酒学硕士学位（他的妻子是他的同学，在继续深造后成为酿酒学讲师，她的三个学生都加入了夏桐酒庄）。艾伦曾访问过法国、意大利、阿根廷、巴西和美国的加利福尼亚。纳帕的"优异风土、气候和顶级葡萄酒"给他留下了深刻的印象，他以极大的热情谈到加州成熟的葡萄酒旅游业。

他用流畅的英语介绍了夏桐酒庄从一开始就生产的干型起泡酒和较新的被称为"蜜"的酒款（"蜜"在中文里是"蜂蜜"或"甜"的意思），他解释说后者的创作目标是生产一款具有更多果实的甜感和柔和的酸度的易饮型起泡酒，专门针对中国市场。尽管中国的起泡酒消费量仅有静止葡萄酒的百分之一，但艾伦坚信起泡酒在他的祖国会有前景："中国的年轻一代喜欢新事物，他们的饮酒习惯更加开放。起泡酒代表了这些新消费者所追求的生活方式。"

时间会证明他是否正确，但是他的信念已经得到了《葡萄酒情报》（*Wine Intelligence*）研究的证实，所以艾伦还是很有机会的。

杨伟明

杨伟明，32 岁，他有着瘦削的身材，修长而优雅的手指，说话温柔，精通英语和法语。他是志辉源石酒庄的酿酒师。志辉源石是一家令人震撼的酒庄，是用当地的石头建造起来的，目的在于吸引游客和葡萄酒专业人士（酒庄拥有迷人的建筑和花园，长而明亮的品酒桌，精致舒适的住宿条件，宽敞、装饰质朴别致而又与众不同的餐厅和一个葡萄酒商店）。

在位于法国叙兹拉鲁斯（Suze-la-Rousse）的葡萄酒大学完成农业工程学位和一年制侍酒师课程后，酿酒师杨伟明（他的英文名是乔纳森）于 2009 年加入志辉源石酒庄，当时建筑工作刚刚完成（该酒庄的第一个年份是 2010 年）。

当他准备用庄园 809 公顷（2 000 英亩）葡萄园生产的葡萄酿造第 9 个年份的葡萄酒时，他评述说："一开始我们想酿造类似波尔多的葡萄酒，因为关于波尔多的话题很多。但是随着岁月的流逝，我们已朝着展示贺兰山风土的风格发展。我们想酿造出适合中国口味的葡萄酒：这是我们未来的目标。"

他意识到这可能很困难："在我的工作中，最大的挑战是怎样在了解宁夏的风土和为中国消费者酿造葡萄酒之间建立联系。我花了很多年时间，使用不同的技术和酿酒辅料来充分表达我们所生产的葡萄的潜力，但仍有许多工作要做。"

他在法国的经历使他对罗纳河谷的葡萄酒情有独钟（他提到教皇新堡和瓦奇哈斯带给他灵感）。他执着于为酿造优质葡萄酒而奋斗——"越努力尝试，面临的挑战就越多，所以我永远不会感到无聊"——尽管中国市场对橡木桶陈酿的红葡萄酒有明显的偏爱，但杨伟明却酿造出了细腻、不经橡木桶陈酿的霞多丽。他谨慎地说："我们这个地区的葡萄酒需要更多的时间来发展，首先要找准其在国内市场的定位，然后才是海外市场。我们必须抓住一切机会，使我们的葡萄酒在宁夏以外广为人知，并且欢迎一切想要来这里发现宁夏葡萄酒的游客。"

中国日报美国版
宁夏葡萄酒惊艳联合国

原文链接：

usa.chinadaily.com.cn/a/201811/09/WS5be46b6ca310eff3032877ca.html

作者：肖红

2018年11月9日

本周，他们在纽约联合国总部举杯致意宁夏葡萄酒。

作为本周星期一至星期五举行的中国美食节的一部分，星期三的专场午餐会有20多位驻联合国大使和外交官参加。

午餐会的美食由来自宁夏回族自治区的厨师准备，特色是来自中国西北贺兰山东麓产区的曾获国际奖项的葡萄酒。

乌克兰常驻联合国代表团顾问拉维·巴特拉（Ravi Batra）在午餐会上分享了一则有关中国白酒的故事。他回忆起有一次印度大使接待中国外交官，端上的是中国最著名白酒——茅台。

"我觉得中国不再会有什么比茅台酒更打动我的了，但是今天，它（宁夏葡萄酒）改变了我的看法；宁夏葡萄酒真的太美好了！"他品尝了葡萄酒后由衷地赞叹道。

宁夏葡萄产业发展局局长曹凯龙对客人说，这些酒是用位于北纬37°～39°的宁夏贺兰山东麓种植的葡萄酿造的。

他补充说："这里被认为是世界酿酒葡萄种植的'黄金地带'。"

曹凯龙说："土壤、阳光、温度、降水、海拔、水热协同效率和其他条件构成了完美的结合，从而生产出香气饱满、颜色美丽、酸糖比均衡的葡萄。"

宁夏贺兰山的葡萄酒产区拥有86个酒庄，种植了近4万公顷的酿酒葡萄，每年可生产大约10万吨葡萄酒。这也是中国唯一的一个酒庄酒产区。

曹凯龙自信地说："它已经成为中国最有前途的葡萄酒产区，与世界上最好的葡萄酒产区不相上下。"

摩尔多瓦前常驻联合国代表亚历山德鲁·库巴（Alexandru Cujba）回忆起他去年9月对宁夏产区的访问。

他说："我在那里的所见所闻给我留下了深刻的印象。这个产区出类拔萃。宁夏在农业领域，尤其是在葡萄和葡萄酒产业上，成就非凡。"

格鲁吉亚常驻联合国代表卡哈·伊姆纳德泽（Kaha Imnadze）表示："全球对葡萄酒的认识正在觉醒，中国就是一个例子，越来越多的人进入葡萄酒（酿造）行业。"

伊姆纳德泽说："大多数行业都存在竞争，但在葡萄酒行业，进入的人越多，对每个酿酒国家来说都越好。"

宁夏作为葡萄酒产区的最大优势在于其葡萄品种和巨大的潜力，这使宁夏正在成为一个全球知名的葡萄酒产区。

人们倾向于寻找新的葡萄酒，因为他们想品尝不同的东西。不一定非得是名头很大的葡萄酒，而是来自酒庄的优质葡萄酒。

伊姆纳泽曾多次作为游客或政府客人到过中国，他说他看到中国在许多领域所取得的成功。

伊姆纳泽说："我的国家格鲁吉亚是葡萄酒的摇篮，拥有500多个葡萄品种，但不是所有的品种都有充分的商业化种植。"

他邀请宁夏代表团访问格鲁吉亚，以促进两国在葡萄种植和酿酒领域的交流与合作。

他说："宁夏可以成为葡萄酒生产的领导者，为世界提供一些许多人都不知道的最新口味。"

2018年中国美食节由人类健康组织、宁夏葡萄产业发展局和联合国使团餐厅共同组织。

美食节以正宗的中国菜为特色，融合了中国传统风味，例如宫廷菜和来自宁夏的地区性美食。

饮料商务
宁夏葡萄酒列车鸣笛发车

原文链接：

thedrinksbusiness.com/2018/08/china-launches-ningxiawww.thedrinksbusiness.com-wine-train-from-beijing/

作者：娜塔莉·王

2018年8月21日

中国推出了一趟连接首都北京和位于西北部宁夏葡萄酒产区的葡萄酒列车，旅客将有机会品尝葡萄酒，尝试葡萄酒面膜，并在酒庄获得第一手采收经验。（节选）

Just-Drinks
为什么世界葡萄酒商不能在中国为所欲为

原文链接：

just-drinks.com/comment/why-the-worlds-wine-producers-wont-have-it-all-their-own-way-in-china-comment_id126945.aspx

作者：克里斯·洛什

2018 年 10 月 11 日

大多数成熟的葡萄酒生产商似乎都认为，中国人会坐以待毙，看着他们的葡萄酒充斥中国的货架并从中获益，尤其是每瓶 100 元人民币（14.5 美元）以上的价格区间。然而，宁夏产区的崛起，证明了法国、澳大利亚等国必须激烈地争夺市场。

2019

Wine-Searcher
中国葡萄酒破壳而出

原文链接：
wine-searcher.com/m/2019/04/chinas-wine-egg-hatches

作者：吉姆·博伊斯
2019年4月28日

宁夏一个新酒庄的发展令人印象深刻，但吉姆·博伊斯指出，这是一场赌博。

在中国西北部宁夏尘土飞扬的土地上，一座被称为西鸽的蛋形建筑群拔地而起，这片地区在英文中被非正式地称为鸽子山。在过去的两年里，住在这片被大风所吹刮着的平原上的居民目睹了从令人鼓舞的蓝图到一个坐落在弧形石墙内的庞大酒庄的全过程。现在，第一批葡萄酒即将上市。

但是西鸽酒庄并不仅仅是又一家高档酒庄。宁夏地区的政府希望它能成为类似"奔富"（Penfolds）一样的存在，一个受人尊敬的高产量的民族品牌，生产从适饮的入门级葡萄酒，到最精致的珍藏葡萄酒。而且，通过这样做，还可以帮助葡萄酒行业的其他人，因为物以类聚，对吧？

该项目暴露了宁夏和中国葡萄酒行业面临的挑战。在不到十年的时间里，宁夏从一个几乎默默无闻的地方，变成了一个充满抱负的名地，赢得了1 000多枚奖牌和来自世界顶级评论家的赞誉。但这种在评论领域的成功并没有在销售上得到充分的反映，这对政府和葡萄酒行业都意味着压力。

公平地说，这不全是宁夏的错。早在该地区加入葡萄酒市场之前，大型生产商就利用消费者的无知，强调营销而非质量，为一场信誉危机埋下了隐患。当葡萄酒质量终于开始上升时，许多对葡萄酒略知一二的消费者开始对本土品牌嗤之以鼻，转而青睐进口葡萄酒。终于落得个自作自受。

一个2.5万米^{2}的酒庄能帮他们打败这场信任危机吗？

事实上，西鸽有卖点和优势。它有着宁夏最大、最老的葡萄园，占地约1 000公顷，绝大部分是赤霞珠，可以追溯到20世纪90年代。他们还有300公顷的新葡萄园，种植着马尔贝克和马瑟兰等品种。

这些葡萄被送到一个总容量为1 000万升的现代化加工车间，里面配备了一流的设备，从德国产的压榨机，到新西兰的VinWizard酒罐控制系统，到成千的仅用于2017年份的橡木桶。

还有一个由当地和国际专家组成的团队管理使用着这些设备。

这里的关键是酒庄老板张言志，他既是在波尔多受过培训的酿酒师，又是从事进口和葡萄酒分销的酒易酩庄公司的负责人，构成了双重因素。酒易酩庄已经经营了一些重要的品牌。这个已有的分销网络很可能是西鸽的成功要素，因为事实证明，对其他人来说很难找到进入市场的渠道。张言志也得了资金支持；据当地媒体报道，西鸽起步获得了4 000万美元的资金，这笔钱来自宁夏葡萄酒交易博览中心的支持。

张言志的团队包括首席酿酒师廖祖宋，他曾在中国最好的酒庄之一怡园酒庄工作，并在澳大利亚的贝斯菲利普（Bass Philip）和莫利杜克（Mollydooker）工作过。此外，还有布鲁诺·威滕内斯（Bruno Vuitennez）和瓦莱丽·拉维尼 Valerie Lavigne 加盟担任公司顾问，克里斯泰尔·切内（Christelle Chene）也加入担任出口销售总监。

万事俱备，第一批葡萄酒就像鸽子排成一排，即将推出。最先面世的三个等级葡萄酒包括入门级的 N28。它也被称为贺兰红，以宁夏的山脉命名，是政府背书的代表该产区的葡萄酒之一。如果你参加官方活动，你可能会喝到几杯。考虑到与奔富的对应，N28 不完全是奔富洛神山庄（Rawson's Retreat），它的零售价定为 24 美元，我听说将会推出更便宜的选择。

再往上就是 N50 老藤葡萄酒，酿酒的果实来自那些有 20 多年历史的葡萄园。再往上是玉鸽系列，包括单一园霞多丽、赤霞珠和蛇龙珠。

最高等级，西鸽珍藏葡萄酒，还在木桶中陈酿成熟。

大多数葡萄酒应该会被投到酒易酷庄的营销网络中，其中一些将流向企业和希望出口的客户。

对了，西鸽的酒店应该在几周内就能对外开张了，这样客人们就能轻松地畅饮美酒，然后再好好地睡上一觉。

西鸽酒庄雄心壮志，但这也带来了风险。如果这些高端设备、葡萄园和顾问，加上政府、金融和经销商的大力支持，都不能实现各项预期呢？然后会怎样？如果这些葡萄酒取得了成功，但西鸽酒庄的巨大规模削弱了众多的小酒庄，而不是促进了它们的发展，又该怎么办呢？

所有的这些都让西鸽酒庄在某种程度上成了一场赌博。话又说回来，很多宁夏葡萄酒的发展从这个角度看又何尝不是如此呢。

无论如何，时间的车轮都在转动。今年 3 月，西鸽的酒样在成都举行的盛大的年度糖酒会上亮相。在李志延和马会勤的带领下，西鸽酒庄还参加了宁夏葡萄酒在多个城市的推介。酒庄接待了源源不断的访客，其中包括许多官员，它即将成为中国葡萄酒的中心舞台。

有趣的是，西鸽并不在著名的贺兰山区域内，贺兰山是该地区那些最著名的酒庄所在地，如银色高地、留世、贺兰晴雪、迦南美地和夏桐。西鸽坐落在更南的地方，处于山脊的末端，那里的风更强，冬天也更严酷。

这里的葡萄藤有着独特的历史，独树一帜。如果它们能说话，可能会讲述自己曾经是与保乐力加合资项目的一部分，以及最近为有来自 17 个国家 48 名选手参加的宁夏酿酒师挑战赛（10 万美元奖金）提供过葡萄。

还有一个深深扎根于过去的故事。1997 年，中国国家主席习近平当时作为福建省的官员访问宁夏，执行扶贫任务。一项成果是把贫困群众从宁夏环境更为恶劣的地区转移安置到了离西鸽不远的地方。在这些地区兴起的一个行业就是葡萄酒。事实上，西鸽的第一批葡萄籐就是在习近平访问的同年种植的。那些相信命运的人可能会认为这是决定项目成功的另一个重要因素。

饮酒杂志
龙的秘密——宁夏如何将中国葡萄酒标上世界地图

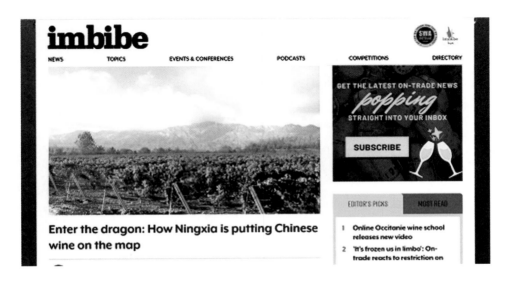

原文链接：

imbibe.com/news/enter-the-dragon-how-ningxia-is-putting-chinese-wine-on-the-map/

作者：克里斯·洛斯

2019年6月14日

宁夏在十年多的时间里就确立了自己作为中国顶级精品葡萄酒产区的地位。克里斯·洛斯前往贺兰山东麓，看一看这个快速发展的产区是否真的能成为中国的纳帕。

伦茨·摩塞尔（Lenz Moser）放下折光仪，凝视着眼前刚刚采收过的赤霞珠葡萄园。

"你知道，中国思虑长远，同时又能在短时期内完成目标。"这位奥地利酿酒师说："这个国家变化的速度着实令人惊叹。"

他说得没错。我们正站在银川市边缘，银川距中国首都北京以西两小时的航程。30年前，银川只是一个3万人口的小镇，现在是200万人口的城市，并且政府计划在未来10年内将其人口增加一倍，达到400万人。八车道的快速路四通八达延伸进旷野，这些空旷之地将很快建起高楼大厦。

银川是宁夏最大的城市。如果您需要一个词来了解中国葡萄酒，那这个词便是"宁夏"。"宁夏"靠的不是葡萄酒的产量，其产量仅占中国葡萄酒总产量的10%。但宁夏对优质葡萄酒的专注，使它获得的行业重要性远比产量占比高得多，人们对宁夏产区有一种确定无疑的兴奋之声，预感这里可能是让中国葡萄酒屹立世界葡萄酒地图的地方。

扎根

摩塞尔虽然出生于奥地利，但他在世界各地酿造葡萄酒，包括美国的加利福尼亚，他把这里的精神与干劲比作20世纪70年代的纳帕谷。当时传奇的加州酿酒师罗伯特·蒙大维（Robert Mondavi）周游世界，告诉任何愿意听他说话的人：他的葡萄酒值得与欧洲最好的葡萄酒进行对比评价。如果有人怂恿你说中国葡萄酒是边角旮旯而不值得关注，请回忆一下当年的纳帕。

在与中国最大的葡萄酒公司张裕合作十多年之后，摩塞尔与张裕共同在宁夏建立了一个新酒庄。张裕摩塞尔十五世酒庄是一个仿佛直接从波尔多搬来的香草色的洛可可风格建筑，其雄心有目共睹。

过去3年中，摩塞尔花了很多时间向欧洲各地发送自己的葡萄酒，并成功地拴牢了遍及欧洲大陆的客户和进口商。

"首先，我们必须做也是正在做的，是带给人们一个惊喜。"他说，"然后，我们还要不断提高。"

对摩塞尔而言，2015年是个分水岭。在经历了艰难的2014年之后，对于一个新酒庄来说这是一个不错的年份，同时他的葡萄树到2015年也已经10年了，这意味着它们"开始变得有趣起来"。

突破

年轻的葡萄藤让人想起这里葡萄酒产业正处于多么年轻的状态，发展又是多么的快速，就像这座城市本身一样。这里最早有记录的葡萄酒厂出现在 20 世纪 80 年代中期，爱笑而健谈的张静在二十年后建立了第一家精品酒庄。

访问她的贺兰晴雪酒庄，酒庄名字的含义是"在一个晴天，从这里可以看到贺兰山上闪闪发光的白雪"。酒庄里陈列的照片上有推土机在一片荒漠中整地的情景。这些照片是 2005 年拍摄的，仅仅 6 年之后，张静的加贝兰赤霞珠混酿就在醇鉴国际葡萄酒大赛中获得了"年度国际最佳葡萄酒"奖。

这一国际认可打开了发展之门。宁夏目前已有 3.8 万公顷的葡萄种植面积（与新西兰差不多）和 100 多个酒庄。未来 10 年，葡萄园的面积预计会增加到 6 万公顷。

将这里的变化速度描述为"疯狂"应该是挺贴切的。这里不缺地——宁夏是中国人口分布最少的地区之一，这里到处都是从地里冒出的葡萄藤和酒庄。

这是记者身处于一辆飞驰的汽车，身后是远去的尘土飞扬的观感。

我们可以看见我们想参观的新酒庄贺金樽，但我们却无法抵达。一年前，这个地方尚不存在，所以没人知道方向，目前因为道路建设还没有完成，路标也还没竖立。事实证明，也没有酒庄。

"他们确实在建造酒庄。"和蔼的新西兰酿酒师戴夫·泰尼（Dave Tyney）说道，"两个星期前，这里地板很脏，而且也没有酒罐。现在他们放进了一个酒罐，第二天我们就装上了酒。这种事让人压力山大，但你会逐渐习惯的。"

宁夏和马尔堡是正式的"姐妹地区"，在半建成状态的混乱之中，不乏新西兰人在工作。很难想象地球上还有什么地方比这里更不像马尔堡，在这里，年轻的团队为工作狂灌咖啡，疯狂地在纸上书写，这可能也是吸引力的一部分。宁夏有其粗糙的地方和怪异之处，但不可否认的是它实在令人热血沸腾。

出路还是疯狂

查看地图，你会看到宁夏和纳帕谷处在同一纬度上。但是两者几乎没有其他相似的地方。纳帕全是山谷和海雾，而宁夏本质上是一片高海拔的荒漠。

宁夏的葡萄园坐落在一个巨大、空旷而多尘的平原上，海拔高1 100米，距离最近的海边1 000千米。夏季炎热，冬季严寒，是典型的大陆性气候。在西部高耸的贺兰山脉的阴影下，降雨十分稀少——每年只有200毫米，且大部分降雨发生在7月和8月。

除此之外，这里是葡萄种植的绝佳气候。即使在盛夏，也不会出现难耐的酷热。由于气温很少会超过35℃，所以葡萄藤不会罢工。

这使本文的作者想起了阿根廷门多萨，在门多萨海拔高度造成昼夜温差很大，从而保留了葡萄的天然酸度。在仲夏夜间的温度通常会降至20℃以下，而在接近葡萄采收时温度甚至降到了个位数。

温度的快速下降是宁夏的特点之一。冬季严寒可以到-20℃，这还不包括风寒，并且严寒来得极快。由于葡萄藤在低于-18℃死亡，中国种植者们有一个独特的解决方案来防止这种情况的发生——他们把葡萄藤埋起来。

葡萄采收之后，很快就进行修剪，然后将主干放倒并用泥土覆盖，看起来像一行行的撒克逊人埋葬的土丘。种植者随后对土地进行灌水，使其冻结起来。包裹在冰中的葡萄藤受到了保护，可免受深冬可能致命的超低温的影响。

需要在生长季结束时掩埋葡萄藤影响着宁夏葡萄栽培的全过程。例如，葡萄行的间距必须足够大，使拖拉机能够在行间行驶，犁起土壤对葡萄藤进行覆盖，这意味着不可能实施高密度种植。

传统上葡萄被整成"独龙干"，这使压埋较为容易，但这样做并不总有利于最大限度地提高葡萄的成熟度。压埋也会对葡萄藤造成损失。

由于需要在10月中旬前完成葡萄的采收，在11月初之前开始修剪（以确保在11月20日之前安全埋土），这里的种植者不可能为了多积累一点糖度，让葡萄在藤上多挂几周。

"我们面临的最大挑战是生长季短。"超级漂亮的志辉源石酒庄的酿酒师杨伟明说,"霜冻可能会在9月底降临。如果我们要新鲜,那必须尽早采摘,但是酚类的成熟较晚。"

杨伟明是推动宁夏葡萄酒革命的众多聪明的年轻中国酿酒师之一。他们都具有很高的资历和丰富的国际经验,似乎都在获得学位之后在法国工作了一段时间,许多人还有在欧洲其他国家和新世界(虽然较不普遍)的经历。

他们并不特别热衷于果实的成熟,而是带回了欧洲式的对葡萄酒结构和新鲜度的追求。银色高地的高源在波尔多的凯隆世家工作了数年,带着她的法国丈夫返回了宁夏。

"我把法国风味带给这里,而不仅仅是成熟的葡萄。"她说,"我更喜欢2014年这样不太理想的年份,这些年份的气候较为冷凉,葡萄酒的酒精度较低。"

杨伟明关于酚类成熟度的观点很好。从这些葡萄酒中能找到共同的线索,那就是轻盈的水果,然后是嘎吱作响的单宁,这样的葡萄酒优雅登场,但在酒体的末尾却略带青涩。

这可能部分要归因于酿酒师为追求新鲜度而愿意付出的成本,但这也与葡萄栽培方式有关。尽管每公顷的产量看起来不错,但考虑到行距较远,以有利于冬季埋土防寒,单株的产量实际上都可能相对较高。

同样地,葡萄园管理还不能与酿酒厂中毋庸置疑的专业知识匹配。相应的叶幕管理技术可以减慢成熟速度,使酚类成熟度赶上糖成熟,这类技术很少或几乎没有被实践。

在张裕摩塞尔十五世酒庄,雇用一名本地的葡萄园经理是摩塞尔优先考虑的事情。他说:"我想与中国人一起工作,因为我们在中国!但是在葡萄园里这件事要困难得多。这里的大学将重点放在酿酒上,而不是在葡萄栽培上。"

新曙光

这也许就是宁夏面临其最大挑战而发起的革命。毕竟,几年内就可以学会如何酿造出优质的葡萄酒,但要了解种植什么品种以及如何使葡萄产出最优可能需要上百年的时间。这里不同酒庄对此有不同的观点，比如览翠酒庄的庄主邵青松问道:"我们为什么要复制纳帕或波尔多？"，而伦茨·摩塞尔这样的人则在想"为什么要再重新发明轮子？"

当前可以肯定的是，波尔多的影响力正日益凸显，大多数酒庄都在采用一种基于赤霞珠的酒庄模式，这种模式与当地市场相适应。

赤霞珠在这里显然是一个具有潜力的品种，尽管本记者认为向其中调入少量的美乐，以支撑口感中段并缓解单宁时，可能表现更为优秀。

但你不禁想这个地区的独特性意味着应该还会有另一个品种在这里发迹，并使宁夏像门多萨的马尔贝克或马尔堡的长相思一样，真正扬名天下。

几年前，有很多关于蛇龙珠（佳美娜）的谈论，鉴于它自杀般的晚熟特征，在容易出现秋季霜冻的地区种植它看起来有点自讨苦吃。许多酒庄都有试验性的葡萄园，目前正在试种包括丹魄、增芳德、麝香葡萄、慕合怀特、西拉和雷司令在内的各种品种。但是在与酿酒师的讨论中出现得最多的品种是法国的红葡萄品种马瑟兰，它是法国用赤霞珠与歌海娜杂交的品种，很有潜力，一些酒庄如蒲尚，酿出的马瑟兰获得了高价的青睐。

尽管如此，无论是波尔多混酿，还是马塞兰，都还没有一个明确的"宁夏风格"。这很可能需要更长的葡萄树龄和更好的葡萄栽培知识。现在还处于早期阶段。

那么，在接下来的几年里英国的餐厅是否会充斥着宁夏葡萄酒呢？可能不会。但也有少数优秀的生产商正在努力向国际扩张，他们的葡萄酒确实有优点。加贝兰已经出现在英国市场（通过熊猫葡萄酒），还有张裕摩塞尔十五世酒庄（通过英国酒商 Bibendum），我预计银色高地很快也会出现。

现在才刚刚开始，但正如一句著名的中国谚语所说:"千里之行，始于足下。"

印度葡萄酒学院
第三届世界葡萄酒博览会彰显宁夏的中国顶级葡萄酒产区地位

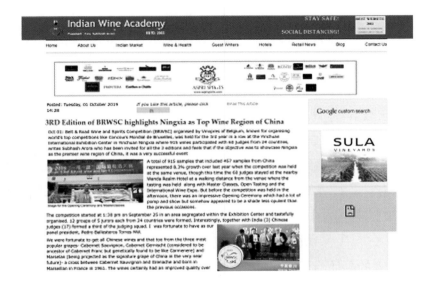

原文链接：

indianwineacademy.com/item_6_824.aspx

作者：Subhash Arora
2019年10月

"一带一路"葡萄酒及烈酒竞赛（BRWSC）由比利时 Vinopres 组织，该组织以主办如布鲁塞尔国际葡萄酒大奖赛（Concours Mondial de Bruxelles）这样的顶级国际比赛而闻名。"一带一路"葡萄酒及烈酒竞赛已经连续 3 年在位于宁夏银川的银川国际会展中心举行。来自 24 个国家的 60 名评委品鉴了总共 915 款葡萄酒。本文的作者 Subhash Arora 连续三次被邀请参加了该项比赛，他认为如果比赛的目的是彰显宁夏作为中国葡萄酒的首善之区的话，这是项非常成功的活动。

在总共915款参赛样品中，有457款来自中国，与去年在相同场地举行的比赛相比有8.3%的增长。这次60名评委都住在附近的万达酒店，步行即可到达品酒现场，这个场地还是葡萄酒大师班、开放品尝和国际葡萄酒博览会的会场。在下午的比赛之前，举行了令人印象深刻的开幕式，有很多华丽的场景和表演，但与以前的开幕式相比似乎少了一点儿奢华。

葡萄酒评比于9月25日下午1:30在展览会中心的一个分区举行，组织安排井井有条。来自24个国家评委分成12个小组，每个小组由5名评委组成。有趣的是，中国评委（17人）加上印度评委（3人）占到了评委队伍的三分之一。我有幸与佩德罗·巴列斯特罗斯·桃乐丝（Pedro Ballesteros Torres）葡萄酒大师分在一组，他是我们小组的主席。

我们很幸运地品尝到了所有的中国葡萄酒，这些葡萄酒由三种最受欢迎的葡萄：赤霞珠、蛇龙珠（被认为是品丽珠的祖先，但发现基因像佳美娜）和马瑟兰（被预测为中国未来标志性的葡萄品种）酿造，马瑟兰是赤霞珠和歌海娜的杂交后代，1961年诞生于法国。与去年在同一地点举办的品鉴相比，这些葡萄酒的品质有明显的提高。

从统计学上看，有17个国家的葡萄酒参加了比赛，其中中国送样最多（457个），其次是摩尔多瓦42个，希腊31个。不出意料，599个样品为红葡萄酒，251个为白葡萄酒，这似乎高于中国消费红、白葡萄酒的比例。在915个样品中，有65个样品是桃红，这个比例也似乎比实际喝的要高——也许这是一个新趋势。这些葡萄酒争夺金奖（95分以上）、大奖和银奖。

该项葡萄酒比赛是中国"一带一路"倡议的延续，2016年亚洲丝绸之路论坛暨品鉴会在北京房山葡萄酒产区首发。在2018年成功举办这项大赛后，由于银川附近拥有中国酿酒葡萄最大的集中栽培面积，且展现了巨大的潜力，2019年9月25日至28日这项活动回归银川再次举办。更重要的是在中央财政的支持下，当地政府部门的营销推广非常出色。他们为本次活动选定的主题是个性、对话与融合。

大师班

在 9 月 26 日举办了数个大师班……其中包括银川市副市长陈康仁关于葡萄酒和经济的中文演讲。随后，亚美尼亚葡萄和葡萄酒基金会执行主任扎拉·穆拉迪扬（Zara Muradyan）就亚美尼亚葡萄酒作了简短的发言，再之后，葡萄酒酿造学家和葡萄酒顾问乔治·萨曼尼什维利（Giorgi Samanishvili）带领大家品尝了来自格鲁吉亚的五款葡萄酒。伯纳德·伯奇（Bernard Burtschy）提供了一个鲜活的例子，证明有机葡萄栽培在中国并不如想象的那么难，而且具有重要性。

最精彩的是李德美教授的演讲，他现在对于任何一个中国研讨会都是熟悉的面孔。作为中国十大葡萄酒顾问之一，他使宁夏产区获得了世界的关注，贺兰山东麓的贺兰晴雪酒庄是他提供咨询的酒庄之一，该酒庄 2009 年份的加贝兰特级珍藏葡萄酒赢得了 2011 年 10 英镑以上最佳国际葡萄酒的大金奖。李德美老师介绍了中国葡萄酒尤其是宁夏葡萄酒，并带领大家对五款葡萄酒进行了品鉴。

酒庄参观和品鉴

在会展中心举办的国际葡萄酒博览会有为期 3 天的开放品酒，来自 20 个国家的 200 多家酒庄在此期间展示了他们的葡萄酒，让参观嘉宾有机会品尝到来自中国特别是宁夏的一些葡萄酒，除此之外，评委们还在繁忙的日程中应邀参观了多个酒庄……

本次活动由布鲁塞尔国际葡萄酒比赛与北京国际葡萄酒和烈酒交易所共同策划协作举办，我们曾在 2016 年参观过北京国际葡萄酒和烈酒交易所……

有一点可以肯定，那就是宁夏在不久的将来会让世界都听到他们雄狮般的声音。中国政府和银川都在为正确的决策下注。宁夏葡萄酒的价格仍然很高，但未来好酒应该匹配以更适当的价格。否则，宁夏葡萄酒的魅力可能会消失，至少在国际市场上会是如此。

贝丹和德梭
贺兰红，中国葡萄酒新品牌

原文链接：

mybettanedesseauve.fr/2019/10/08/helan-hong-la-nouvelle-marque-de-vin-chinoise/

作者：马蒂尔德·于勒
2019 年 10 月 8 日

在中华人民共和国成立七十周年庆祝活动全面展开之际，在习近平总书记的支持下，宁夏回族自治区政府推出了一个名为贺兰红的葡萄酒品牌。一个新的产区在东方冉冉升起，很快就会超过波尔多葡萄酒产区。

"贺兰"指宁夏葡萄园地处的山脉，位于北京以西 1 100 千米之外，"红"指红色，是中国朋友心中幸福与成功的象征。事实上，除了长城、龙徽和张裕之外，还没有一个响亮的名字，能让更多的人知晓这个全新的产区——宁夏。宁夏产区的事业刚刚起步，它的"老葡萄园"只有 20 年的历史，但其发展势如破竹。中央和当地政府已经把它作为中国美食复兴的旗舰葡萄园。它位于戈壁沙漠的边缘，占地4万公顷，几年之后在规模上可能会超过我们古老的波尔多葡萄园。这项事业对中华民族来说，不是一件难事，无论严寒（葡萄必须在冬天埋土，冬天的温度可以降到-30℃）、酷暑、干旱还是萧条的市场，都不会阻挡他们前进的脚步。然而，唯一的难题就是水源。人工开辟的葡萄园面积太大，黄河很难满足其迫切的灌溉需求。

贺兰红，如雷贯耳，在2018年底出现在领导人的脑海中。这是一个公用品牌。以后一定会有多种产品，以贺兰红之名，呈现给大众。目前，此品牌旗下只有赤霞珠葡萄酒。根据文件规定，有四家公司被授权生产，分别是贺金樽、贺兰神、西鸽和御马酒庄。第五个酒庄正在国家的赞助下建造。目前只生产了10万瓶，其中3 000瓶是为我们的比利时朋友准备的，他们正在布鲁塞尔市中心的天堂动物园（Pairi Daiza）游乐场庆祝这款新酒。它清新宜人，是年轻的宁夏产区的最佳代表。

用欧洲的文化比喻来说，贺兰红是理想的特洛伊木马，侵入了我们西方市场。从长远来看，中国计划生产2 000万瓶贺兰红葡萄酒，其中"仅"1%出口。这一新品牌将使宁夏葡萄酒更受欢迎。不过，宁夏葡萄酒每瓶售价25～150欧元，对普通人来说仍有些奢侈。

葡萄圈
夏桐推出其中国首款起泡红葡萄酒

原文链接：

vitisphere.com/actualite-90590-Chandon-lance-son-premier-vin-rouge-petillant-de-Chine.htm

作者：亚历山大·阿贝兰
2019 年 11 月 13 日

去年 9 月，路易·威登－酩悦·轩尼诗（LVMH）集团旗下的葡萄酒部门推出了禧起泡酒（cuvée Xi）。这款酒优先定位于营业中的咖啡馆、酒店和餐厅网络，目标是在中国创造新的消费方式，为起泡酒开辟新的销售途径。

2020

华盛顿邮报
引领中国葡萄酒革命的女性

原文链接：
washingtonpost.com/photography/2020/01/24/these-are-women-leading-chinas-wine-revolution/

作者：马蒂尔德·加托尼、马蒂奥·法古托
2020年1月24日

　　宁夏是中国西北部地区的神秘瑰宝。纵观其雄伟的山峰和蔓延到视野尽头的整齐的葡萄藤，很难想象20多年前，该地区只是一片自给自足的农民所居住的荒芜沙漠。42岁的任艳玲在难得的休息时间坐在她的实验室中回忆说："小时候，我曾经在荒漠上挖洞。我会躲在那儿，和我的朋友们一起玩耍。"

　　锐利而自信，眼神充满个性和穿透力，琳达是领导着中国葡萄酒革命的一群意志坚强、才华横溢的女性庄主、酿酒师和管理人员中的一个。（节选）

饮料商务
宁夏产区继续成长

原文链接：

thedrinksbusiness.com/2020/04/ningxia-region-continues-to-grow/

作者：爱丽丝·梁

2020 年 4 月 9 日

宁夏是中国著名葡萄酒产区之一，在新冠肺炎疫情全球大流行下，当地葡萄酒产业的发展也并未停止。

与世界上许多其他地区一样，宁夏的葡萄酒产业受到了新冠肺炎疫情的严重影响。

预计很多酒庄将面临现金流困难，为此，宁夏农业农村厅和财政厅已经宣布拨款 3 000 万元用于对行业的补贴。

此外，在 2020 年第一季度，青铜峡、红寺堡和贺兰的 6 家企业提交了建设 5 个新酒庄和扩建一个现有酒庄的计划。预计总投资将超过 2.5 亿元人民币，葡萄酒总产量将增加 2 000 吨或更多。

宁夏贺兰山东麓地区由于其理想的风土被认为是大有前途的产区之一，那里拥有充足的日照、沙质的土壤、低降水量和大陆性气候。

最近，位于宁夏贺兰县的贺兰山东麓葡萄产业园管委会透露：截至 2019 年底，宁夏的葡萄种植面积已达到 57 万亩（相当于 3.8 万公顷），占中国酿酒葡萄种植总面积的四分之一，是全国最大酿酒葡萄集中产区。在该地区广泛种植的葡萄品种有马瑟兰、马尔贝克、黑比诺、雷司令和霞多丽。

当地葡萄栽培的一个显著特点是，需要将葡萄埋土防寒，以保护葡萄藤免受冬季可能降至 -25℃ 产生的冻害。而到了 3 月初，有报道说葡萄园就已经开始分步出土并给葡萄藤施肥。这一过程预计在 4 月中旬完成。

目前，全区有 211 家酒庄，年产葡萄酒 1.3 亿瓶，葡萄酒产业的总产值达人民币 260 亿元。一些精品酒庄，如位于银川子产区的贺兰晴雪酒庄和银色高地酒庄，已经成功地获得了世界知名度。

大型国际品牌如路易·威登－酩悦·轩尼诗和保乐力加等也在宁夏投资，成立了夏桐酒庄和贺兰山酒庄。

近年来，该地区一直在积极推动其葡萄酒产业，并创造了 12 万个就业机会。葡萄酒旅游业是另一个发展重点，该地区目前每年已接待超过 60 万名游客。

人民日报
葡萄酒产业使宁夏的贫瘠土地变得繁荣

原文链接：
en.people.cn/n3/2020/0811/c90000-9720281-2.html

作者：严科、王汉超
2020 年 8 月 11 日

贺兰山东麓位于中国西北宁夏回族自治区省会银川郊外，是一片连绵不断的葡萄酒产区，被誉为宁夏回族自治区的"紫色名片"。

从大约 20 年前种下的第一株葡萄苗，到构建完整的产业链和影响力，葡萄酒产业在这片曾经贫瘠的土地上蓬勃发展。

从 1996 年开始，中国把东西部扶贫协作作为国家重大战略。时任中共福建省委副书记的习近平负责福建帮扶宁夏的工作。闽宁合作从此正式启动。

宁夏的西海固地区曾因土地贫瘠、水资源匮乏被联合国列为"最不适宜人居的地方"。幸运的是，1997 年在宁夏首府银川以南的戈壁荒漠上建立了一个移民安置小镇，吸纳了西海固 4 万多居民。为纪念福建（简称闽）与宁夏（简称宁）的紧密合作关系，该镇被命名为闽宁。

脱贫工作与闽宁的建设和移民安置工作同步进行，葡萄酒产业是该镇摆脱贫困的重要支柱。福建企业家认为，闽宁所处的戈壁滩并不是一般的戈壁滩。闽宁位于地球北纬 38.5°，降水量少，日照时间长，气温日变化大。此外，高渗透性沙质土壤中丰富的矿物质为酿酒葡萄的生长提供了极佳的条件。

随着资金和技术的投入，闽宁的戈壁荒漠很快被绿色覆盖，弥漫着葡萄酒的香气。得益于葡萄酒产业的发展，西海固移民就业稳定，收入不断增加。最后，他们留在了闽宁。

杨成在移居闽宁前在山里生活了几十年，在那里种土豆和小麦，整天为灌溉发愁。自从他搬到新城镇后，他的家人都开始在葡萄园工作。经过培训，杨成了一名电工，他的妻子成了一名环卫工人。另外，他的儿子在葡萄园作挖掘技师。一家三口每月的收入超过 1 万元（约合 1 435 美元）。

杨家人工作的葡萄园由福建商人陈启德经营。从十万亩的荒地开始，陈发誓要酿造宁夏最好的葡萄酒。现在，他酿造的葡萄酒几乎每年都会赢得国际奖项。

截至 2019 年底，贺兰山东麓共有葡萄园 38 万公顷，为搬迁移民提供了约 12 万个就业岗位。

闽宁扶贫从酿酒开始，但从未止步于此。闽宁合作还包括构建可靠的销售渠道和营销方式。

赖有为是福建德化市指定在闽宁挂职的干部。到了宁夏后，他邀请了很多老家的商人到闽宁的葡萄园，向他们推荐产品，拓展销售渠道。今年 5 月，他还参加了一场直播营销，帮助销售葡萄酒，期间他完成了近 30 万元的订单。

来自福建晋江的朱文璋是一位葡萄酒经销商，现在贺兰山东麓工作。基于他与喜欢高品质葡萄酒的酒商的联系，他引入了一种创新的"共享酒庄"的方法。这种方法帮助 50 多家企业"认领"了总计 200 公顷的葡萄园作为其原料来源，使质量得到更好的控制。同时，葡萄种植者也不必担心销路了。

葡萄酒产业见证了闽宁合作 20 年来的努力和成果，帮助许多家庭实现了致富的梦想。如今，闽宁合作仍在升级换代，力争实现更加辉煌的目标。

北京评论
东西部对口合作促进贫困地区发展

原文链接：
bjreview.com.cn/Nation/202008/t20200810_800217158.html

作者：陆燕
2020 年 8 月 13 日

最近，31 岁的张俊宁（音译）买了辆汽车。这是 6 年前难以想象的，那时他和他的家人生活还是很贫困。

那时，他刚从中国西北部宁夏回族自治区的贫困山区西海固搬到约 300 千米外的闽宁镇。

离开了以前的传统农耕生活，他一开始在新环境中茫然无措。他所能做的都是零碎的工作，例如在建筑工地上打工，刚刚够维持生计。

但是当他和他的妻子被一个葡萄园雇用时，压力很快就减轻了。在自学与接受培训之后，他们逐渐掌握了葡萄种植技能，开始获得稳定靠谱的收入。

看到越来越多的像他这样的村民被葡萄园雇用，张俊宁产生了用他所赚的钱建立一个劳务中介的想法，帮助葡萄园雇用当地村民。

他的企业家精神使他不仅摆脱了贫困，而且还盖了一栋新房子过上了舒适的生活。

张俊宁对中央电视台说："如果没有国家脱贫计划，我可能还是个贫穷的农民，不可能实现今天这一切。"

合作的成果

自20世纪90年代中期以来，中国较发达的东部地区和较不发达的西部地区之间的配对合作一直是脱贫和缩小不同地区贫富差距战略的重要组成部分。

在1996年5月国务院召开的一次扶贫合作会议上，中国东部的10个发达省份与西部的10个欠发达地区配对。中国东南沿海的福建被选为援助宁夏的省份。

在宁夏，像西海固这样的一些地区由于极端恶劣的气候条件而处于贫困中。当地人对这个地区的描述是："夏季干旱，秋季洪水，冬季寒冷彻骨。"在1972年，联合国将西海固列为世界上最不适宜居住的地方之一。

在福建与宁夏的伙伴关系下，提出了一项移民转移安置计划，将整个村庄的居民从西海固等贫困地区迁移到靠近黄河的土地更肥沃的地方。闽宁就是转移安置地点之一。

作为脱贫计划的一部分，闽宁的发展始于1997年时任中共福建省委副书记的习近平同志的想法，他当时负责福建协助宁夏的工作。

闽宁这个名字代表福建与宁夏之间的伙伴关系，闽代表福建，而宁则代表宁夏。

现在，闽宁居民超过6万人。根据该镇党委副书记赵超的说法，去年该镇居民的人均可支配收入增长了20倍以上，接近14 000元人民币（约合2 013美元）。

赵超说，由于2020年是实现消除绝对贫困和在各个方面建设小康社会的收官之年，该镇将确保目前剩余的共22个贫困家庭摆脱贫困。

不想搬到闽宁的村民也没有被落下。当地政府支持发展以果树种植为代表的现代特色产业，这既有利于经济发展，也有利于生态建设。这些行业为村民提供了越来越多的工作和其他机会，使他们在西海固过上更好的生活。

人的力量

来自福建的许多政府官员、教师、医务工作者、教授、专家、企业家和志愿者为宁夏的脱贫做出了贡献。

1998年，来自福建的蘑菇种植专家和技术人员来到了当时还处于起步阶段的村庄闽宁。他们为村民提供了从头开始建造数百个蘑菇温室的指导。

其他经济作物也按照类似的方式发展起来。随着种植和养殖业的发展，以及移民人数的增加，该地区出现了更多的村庄。2001年，成立了闽宁镇。今天它下辖6个行政村。

葡萄酒业也是闽宁的支柱产业，闽宁的自然条件适宜葡萄酒的发展。在20世纪90年代末，只有少数几个农民种植葡萄并在家庭作坊里酿造葡萄酒，规模不大。后来，在闽宁伙伴关系的推动下，在福建商人的推动下，葡萄酒行业的规模扩大了。

陈德启就是其中之一。他于2007年访问该地区后，决定在闽宁建立葡萄园。他的投资已变成拥有近500万株葡萄藤和3 000公顷土地的有机葡萄工业园。他的葡萄园雇用了3 000多名村民。

陈德启说："这仅仅是开始。"他补充说，他计划将自己的工业园区规模扩大一倍，为当地居民提供超过1万个就业机会。

截至2019年底，福建共有5 700家企业参与了宁夏的产业发展，涉及20多个行业，年产值数百亿元。

闽宁伙伴关系不仅限于经济发展。在过去的24年中，福建的1 000多名教师和260多名大学生前往宁夏的中小学支教。福建在宁夏建立了236所学校，援助了9万多名贫困学生。

福建省政府还参与了 300 多项在宁夏的医疗卫生计划，包括建立了母婴护理服务中心和医疗培训机构。

福建已有 180 多名官员在宁夏挂职，开展脱贫工作。

宁夏扶贫办负责人梁积裕说，从福建来宁夏工作的同志不仅带来了业务、资金和技术，还带来了先进的发展理念和丰富的经验。

福建莆田扶贫办官员胡家悦（音译）说："福建和宁夏虽然相距千里，但都怀有相同的发展愿望。这两个地区将加强互助和相互学习，并开始新一轮的合作。"

新华社
中国最重要葡萄酒产区的远大梦想

原文链接：

xinhuanet.com/english/2020-10-24/c_139464842.htm

时间：2020 年 10 月 24 日

于本周五结束的国际葡萄酒博览会再次使中国宁夏回族自治区成为焦点。

在贺兰山东麓举办的第九届宁夏国际葡萄酒博览会为期两天，博览会吸引了来自 96 个国家和地区的数百名专家参加线上和线下活动，并就葡萄酒行业的发展发表意见。

总部位于巴黎的国际葡萄与葡萄酒组织主席雷吉娜·万德林娜（Regina Vanderlinde）表示，在扶持葡萄酒行业的优惠政策下，宁夏葡萄酒的质量得到提高，品牌得到打磨，取得了巨大成就。

多年来，宁夏正迅速转型为中国主要的葡萄酒产区。它目前有 3.28 万公顷的葡萄园，年产1.3亿瓶葡萄酒，综合产值261亿元（约合39亿美元）。

宁夏志向高远，计划到2025年将葡萄种植面积翻一番，年产量达到3亿瓶葡萄酒。

戈壁荒漠中的葡萄园

由于阳光充足，灌溉条件优越，气候适宜，宁夏广袤贫瘠的戈壁滩被当作是种植酿酒葡萄的"黄金地带"。

时间追溯到1984年，俞惠明与7位年轻酿酒师，用了4个月的时间，在100多个泡菜坛中酿造葡萄酒。

当时，由于缺乏行业标准，很多中国酒厂倒闭，了解葡萄酒的顾客寥寥无几。俞惠明工作的酒庄，也是宁夏的首个酒庄，也经历了一段艰难的时期。"工人们不能按时领到工资，我的6个同事为了更好的生活离开了。"俞惠明回忆说。

然而，俞选择留在这个行业。

20世纪90年代，一家外国公司收购了该酒庄积压的葡萄酒，然后贴上自己的标签，以每瓶268元的价格出售。

"我意识到我们的葡萄酒有市场，但我们需要自己的品牌来打入市场。"俞惠明说。

2011年，宁夏推出一系列受欢迎的政策，以规范葡萄酒行业，鼓励酒庄生产优质酒和品牌酒。

此后，优惠政策吸引了大量国内外的酒庄在这个地区建立酒庄。于是，一批批优质宁夏葡萄酒流入了市场。

到目前为止，来自50个当地酒庄的1 000多款葡萄酒在世界各地的顶级葡萄酒比赛中获奖。宁夏葡萄酒已出口到20多个国家和地区。

酿酒师的梦想之地

今年35岁的廖祖宋是宁夏西鸽酒庄的首席酿酒师，2014年从澳大利亚的一个酒庄归来。他参与了整个酿酒过程——从西鸽葡萄园和酒庄的建设，到酿酒技术的设计。

"我想做一名了解从葡萄种植到酿造的整个生产环节的酿酒师",廖祖宋说。

西鸽目前拥有超过1 333公顷的葡萄园,年产葡萄酒超过100万瓶。"我相信我们能为全国乃至全世界的客户生产出最好的葡萄酒",廖祖宋说。

同时,为了打磨本土葡萄酒品牌,提升酿酒技艺,宁夏还邀请了世界著名葡萄酒产区的酿酒师。到目前为止,已有来自23个国家和地区的60名酿酒师来到宁夏。

蓬勃发展的葡萄酒旅游

来自北京的游客赵亮(音译)也是一位葡萄酒爱好者,他和朋友一路开车来到西鸽酒庄。他说:"我去过很多国外的酒庄,但宁夏葡萄酒的口感和当地酒庄先进的生产线还是让我很惊讶。"

"我们酒庄离许多旅游目的地都不远,特别适合发展葡萄酒旅游,"西鸽酒庄的创始人张言志说。

宁夏一直在挖掘葡萄酒旅游市场的潜力。目前,宁夏的酒庄每年接待超过60万人次的游客。

贺兰山东麓葡萄产业园的官员赵世华说:"葡萄酒驱动的旅游业对促进该地区旅游业发展起着关键作用,同时也为葡萄酒消费提供了动力。"

南华早报
为何中国的生物动力葡萄酒既能畅销国内，又能受到欧洲和日本鉴赏家的喜爱

原文链接：
scmp.com/magazines/style/luxury/article/3044436/why-chinas-biodynamic-wines-are-gaining-cult-following-home

作者：周慧晨
2020年1月3日

酿酒师们汲取了中国"二十四节气"的智慧，这是一种古老的历法，即使在今天仍然运用于农业安排，酿酒师们发现这是实践生物动力农业的一个很好的参考，因为这两个系统有很多共同点。

"一本出版于6世纪的中文书对决定何时播种、修剪、葡萄园采收或品尝葡萄酒确实方便实用，就像生物动力学日历中的指示一样。遵循这些原则，

就能把葡萄园管理得井井有条",银色高地酒庄的高源解释道。

随着宁夏越来越多的酒庄转向有机农业,生物动力葡萄酒也越来越受欢迎,博纳佰馥酒庄吸引了大量的中国消费者,其中主要是受过良好教育的中年男性。(节选)

环球邮报
首次中国葡萄酒品鉴展现巨大潜力

原文链接：

theglobeandmail.com/life/food-and-wine/article-the-first-real-taste-of-chinese-wine-shows-tremendous-potential/

作者：克里斯托弗·沃特斯
2020年4月24日

　　加拿大安大略省酒类控制局放行了张裕摩塞尔十五世酒庄的4款葡萄酒，张裕摩塞尔十五世酒庄是位于中国北方的宁夏贺兰山东麓的一家精品酒庄，是中国葡萄酒繁荣的第一批惊鸿一瞥。

　　宁夏对品质的重视让许多专家相信，宁夏将使中国葡萄酒与世界知名产区并驾齐驱。类似的情况也曾出现在纳帕谷，纳帕葡萄酒仅占加州葡萄酒总产量的4%，却被普遍视为加州最好的葡萄酒。（节选）

福布斯
宁夏葡萄酒亮相墨西哥城

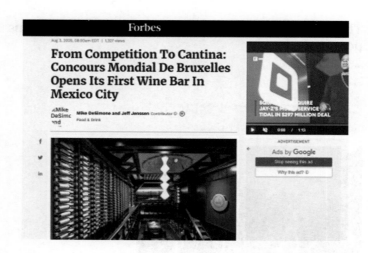

原文链接：

forbes.com/sites/theworldwineguys/2020/ 08/03/from-competition-to-cantina-concours-mondial-de-bruxelles-opens-its-first-wine-bar-in-mexico-city/

作者：迈克·德西蒙和杰夫·詹森
2020年8月3日

布鲁塞尔国际葡萄酒大赛开设葡萄酒吧的构想是为来自世界各地的葡萄酒和烈酒生产商提供一个场所，可以与葡萄酒和烈酒爱好者以及行业从业者，诸如：侍酒师、分销商和进口商等进行交流与互动。

在这个葡萄酒吧的众多葡萄酒产区中，有宁夏葡萄酒的一席之地，这是拉丁美洲首个宁夏葡萄酒亭，它有正宗的中国风格装饰，并提供来自中国宁夏的获奖葡萄酒。（节选）

中国日报
宁夏入选全球十大最具发展潜力的葡萄酒旅游目的地

原文链接：

global.chinadaily.com.cn/a/202010/23/WS5f924640a31024ad0ba80897.html

作者：李映雪（音）

2020 年 10 月 10 日

"世界十大最具发展潜力葡萄酒旅游目的地"榜单揭晓，宁夏入选热门葡萄酒旅游胜地。

其他上榜的旅游目的地还包括智利的迈坡（Maipo）、新西兰的怀拉拉帕（Wairarapa）、加拿大的魁北克（Quebec）和日本的山梨县（Yamanashi-ken）。（节选）

英国广播公司新闻频道
中国消费者对国产葡萄酒日渐喜爱

原文链接：

bbc.com/news/business-55226201China's drinkers develop taste for home-grown wines

作者：蒂姆·麦克唐纳

2020年12月17日

　　高源在法国波尔多地区学习过葡萄酒酿造，她同意这样的说法，如果说消费者转向喜爱中国葡萄酒的话，原因是中国有了更高的标准，而生产商也了解他们的市场。

　　"我相信中国葡萄酒的质量正在不断提高，"她说，"与此同时，新一代的葡萄酒爱好者也更加具有探索与尝试精神，他们为自己国家的传统感到自豪。"（节选）

2021

葡萄酒评论
宁夏，征服沙漠的葡萄酒

原文链接：

larvf.com/,vins-chine-grace-vineyard-ningxia-vignes-wineries,13186,4246418.asp

作者：杰罗姆·鲍杜因
2021 年 2 月 2 日

第 5 天。我们的特使杰罗姆·鲍杜因（Jérôme Baudouin）来到了中国的宁夏地区。仅仅十年，这里的沙漠地区就见证了葡萄的诞生！

我们离开怡园酒庄（Grace Vineyard），前往约一小时车程的太原机场。沿途欣赏周围干旱的气候和风景，我才意识到葡萄树在这里是多么罕见。怡园酒庄周围一望无际的黄土高原上种植了数百公顷的葡萄。但除此以外，什么都没有。实际上，该地区只是山西中心的一个岛状地带。来自该地区的富商陈进强可谓疯狂，他发家致富之后，就想建立自己的葡萄园。

目的地宁夏的情况截然不同。我们向西走，越过鄂尔多斯沙漠后，到达穆斯林聚集的回族地区，这里的葡萄种植已有数百年历史了。宁夏是一个小省，东南方是黄河冲刷而成的沙化草原，西北方是贺兰山山脉。贺兰山本身是抵抗腾格里沙漠的天然屏障，而腾格里沙漠向北延伸至贺兰山山脉以北，是进入内蒙古的标志。只有黄河流过这片土地，作物才得以生长。

第一次参观青铜峡的葡萄园，我就了解了葡萄以及其他农作物的种植方式。一排排的葡萄生长在稻田和玉米地之间。实际上，为了从黄河中引水，这里已经建立起了庞大的灌溉网络。宁夏回族自治区农业厅的一位农学家在建设中的大楼前，向我们解释了这一战略选择的重要性："要知道，十年前，这里除了沙子什么都没有。而如今，我们已经耕种了数万公顷的土地，也借助充足的劳动力，开展了农业活动。"

今天，葡萄酒业被认为是具有很高附加值的战略产业，因此，私人投资者和政府的热情都很高，政府还毫不犹豫地建立了发展援助体系。后来，我们看了一些照片：十几辆推土机排成一排，在干旱的土地上前进，准备修整土地、挖犁沟和灌溉渠。这场战斗似乎已经胜利了，永不干涸的黄河水终于到达了这片土地，而源头就是喜马拉雅山上的巨大冰川。宁夏葡萄园面积已达 24 600 公顷，约有 40 个不同规模的酿酒厂生产葡萄酒。宁夏现在是中国第一优质葡萄酒产区，并于 2012 年 3 月作为观察员加入了国际葡萄与葡萄酒组织（OIV）。此后，宁夏不再只有沙漠。（节选）

爱尔兰时报
走进东方巨龙——包含西方经验的中国葡萄酒

原文链接：

irishtimes.com/life-and-style/food-and-drink/drink/enter-the-dragon-chinese-wines-embodying-lessons-from-the-west-1.4469289

作者：约翰·威尔逊

2021年2月6日

我去年经历的最有趣的线上品酒之一，是与奥地利葡萄酒生产商伦茨·摩塞尔（Lenz Moser）。不过，我们品尝的不是奥地利葡萄酒，而是中国葡萄酒。摩塞尔决定将重点放在中国宁夏的张裕摩塞尔十五世酒庄，在那里他不仅负责葡萄酒的生产，还负责全球销售。

"2015年对我们来说是突破性年份，"他说："由于当年的中秋节比较早，所有采收工都回家去过节了，我们比平常晚了两周收获。当年的酒立刻变得更好，具有更高的酒精度和成熟的单宁，以前的酒精度是12.5%，葡萄酒带有像法国卢瓦尔河产区的品丽珠那样的草本植物特征。"（节选）

詹姆斯·萨克林网站
2020 宁夏报告——快钱 vs 值得陈年的质量

原文链接：

jamessuckling.com/wine-tasting-reports/2020-ningxia-report/

作者：帅泽坤（音）
2021 年 2 月 5 日

宁夏回族自治区位于中国西北的中部，一直以"中国速度"发展其葡萄酒产业。就在五六年前还没有多少酒庄投入运营，而那些已经开始生产葡萄酒的酒庄还在学习生产优质葡萄酒的复杂步骤。

如今，宁夏的酒庄数量已迅速增加到近 100 家，至少还有另外 100 家酒庄在等待政府批准开始生产葡萄酒。宁夏距离北京 1 100 千米，坐落在海拔 1 000 米的开阔高原上。这些酒庄有的端庄，有的奢侈，有的呈现奇异的西方风格，有的具有巧妙的现代感与独特性。它们不再只是光鲜亮丽的门面，从其中许多酒庄出产的葡萄酒现在都是真正的物有所值。虽然在葡萄酒的酿造中仍然存在一些问题，但通往伟大葡萄酒的道路已经铺好。

这是我们第一次对宁夏的近距离考察，甄选了来自该地区的约130款葡萄酒，其中包括一些在酒庄品鉴的葡萄酒：贺兰晴雪、博纳佰馥、留世、立兰酒庄、美贺庄园、西鸽酒庄和银色高地。

绝大多数生产商都按要求发来了样品（包括少量购买的样品），因为他们的酒已经为我们所知。尽管使用软木塞的葡萄酒数量相对较高，令人失望的情况发生很少，总共只有6瓶酒存在软木塞污染。考虑到现在几乎有三分之一的葡萄酒使用DIAM塞，软木塞的污染率非常高，这在宁夏是一个值得注意的问题。如闻酒庄的酿酒师顾嘉威解释说，从葡萄牙和西班牙出口到中国的天然软木塞往往是其质量较低的产品，这就是为什么在过去几年中许多严谨的生产商都转向DIAM塞的原因。如闻酒庄现在正在尝试各种不同封口产品，其中包括一种也能隔绝三氯乙酸的竹制产品。

在品尝的130款样品中，许多酒的得分在88～90分的舒适区间，但只有13款葡萄酒达到93分或更高。这些葡萄酒具有真正的个性，而不仅仅是宁夏红葡萄酒中常见的令人愉悦和口感醇厚（尤其是甜度）。

有人会说"甜水果"就是宁夏红葡萄酒的精髓，这与其气候和风土有关。当然，宁夏的气候赋予赤霞珠和美乐等葡萄品种饱满而富于果味的特征，但我不禁思索，这是否只是没有做到充分尝试或为了满足市场期望而晚采的借口，这些因素比风土和气候对风格的影响更大。毕竟，这份报告中的顶级葡萄酒不存在那种甜得发腻的特性。尽管如此，在品尝了130款葡萄酒后，宁夏没有辜负人们对一个年复一年地获得了无数奖牌的产区的预期。但是，如果想不懈追求严谨的品质，希冀未来有更多的成就，酒庄们仍有很多工作要做。

如今，宁夏生产的葡萄酒酒体坚实、浓郁而又令人陶醉，具有"讨人喜欢"的甜味。不可否认，果香的优势被柔顺的口感所加强，这得益于非常新世界风格的高酒精度（有些酒达到15.5%以上）和甘油，有些葡萄酒还保留了4～5克/升的残糖，使其在面对没有经验的新消费者时更具吸引力。由于赤霞珠在这里不容易充分成熟（当它充分成熟又不过熟时，可以酿造出漂亮的葡萄酒），许多生产商开始晚采，以确保酿出的葡萄酒没有令人不快的青椒特征，这种特征在 10 年前很普遍。当我与生产商交谈时，他们都指出"大多数中国消费者不喜欢酸度和单宁"，而我自己也是中国人，我知道大多数刚开始喝葡萄酒的消费者是这种情况。

但是为了保持葡萄酒的平衡与稳定，许多生产商倾向于在较热的年份对葡萄酒进行调酸。这可能解释了为什么有些葡萄酒尝起来饱满，甚至有果酱的味道，但又有尖酸感，而不是释放出葡萄中多汁的特性。

由于这种风格在市场上广受欢迎，一些生产商就遵循这一套路，而忽略了那些更地道、更复杂的风格。但是，宁夏的顶级生产商理解这个问题，他们倾向于将这类简单、果味型的葡萄酒限制在其入门级的酒款上，以吸引新的消费者。有些人也会避免这种葡萄酒的"糖衣"。张静就是其中之一，她是领先的贺兰晴雪酒庄的联合创始人兼酿酒师。她告诉我，她希望自己的酒具有真正的深度。

张静说："我希望我的葡萄酒在年轻时表现出一定的深度，从而有助于它们的陈年。自 2017 年以来，我们还放弃了调酸，因为我们认为最好保持自然的酸度不变。因为我们一直采摘得比较早，酸对我们来说不是问题。"无论是红葡萄酒还是白葡萄酒，她的葡萄酒都表现出内敛的风格，带有新鲜和微妙感，呈现出高级而精确的感觉，在旧世界的感性中保留了现代性。

同样出自一位女性酿酒师之手，银色高地酒庄生产一系列的葡萄酒彰显酿酒师的直觉。对高源来说，"精确"可能不是最贴切的词。那就是高源和她的父亲高林先生一样，对葡萄酒和这个地区抱有愿景。高源的酒灵魂饱满，拥有本真而又不受拘束的个性，我将这种风格描述为宁夏慷慨的天之馈赠与她酿酒技巧的结合。"就像我们宁夏人一样。"高源轻声说，指的是当地人脚踏实地、好客的个性，如果他们的客人不留下来品尝当地的烤羊肉，他们可能会生气。

如今，高源还希望尽力使葡萄酒地道和自然。为了更加尊重自然，使葡萄藤更健康，具有更强的"免疫"系统，她的葡萄园在 2017 年转而采用生物动力法。高源说："我们一直在努力以更好地理解我们的土地，使葡萄酒拥有灵魂和更少的人为干预。"他们新推出的家园葡萄酒，采用野生酵母发酵和低剂量的硫，展示了她生产品种和产地典型性突出的葡萄酒的志向。

在宁夏什么时候开始采收是十分关键但又棘手的决定，葡萄的生长季不够长，而昼夜温差巨大且光照充足又导致了高糖／酒精含量。因此，成功与失败之间仅差之毫厘。若采收得太早等不到诱人的酚类物质成熟，做出的酒就会有明显的生青味；若采收得太晚，果实过度成熟，潜在的酒精度就会很高，而酸度很低。但是，如果为赤霞珠找到了成熟和新鲜俱佳的平衡点，那么这些宁夏葡萄酒就不仅丰富诱人，而且高级、新鲜还精致。我们酒单上的顶级赤霞珠葡萄酒就表现了这一特点以及对橡木桶的精妙应用。事实上，许多顶级葡萄酒都是橡木桶精选款，比如双瓶装的银色高地艾玛珍藏葡萄酒和2016年贺兰晴雪特级珍藏葡萄酒。

在得到93～95分的顶级葡萄酒中，绝大多数都是赤霞珠或波尔多混酿，主要来自四个顶级酒庄：贺兰晴雪、银色高地、迦南美地和刚成立的嘉地酒园。当我在2016年首次品尝这些葡萄酒时，它们就让我大吃一惊。如今，我认为这四个酒庄在宁夏是行业的领跑者，它们有一个有趣的共同点：都是由女性酿酒师或庄主领导的。其中两家酒庄（嘉地酒园和迦南美地）都由周淑珍做顾问，她是宁夏一位经验丰富且备受尊敬的独立酿酒师，她还担任利思酒庄、留世酒庄和长和翡翠酒庄的顾问。

嘉地酒园信使珍藏2016（95分）是130款葡萄酒中我认为最好的。它展示了如果生产者处置正确，宁夏的赤霞珠可以有多棒。这款成功的2016年份葡萄酒在成熟度和复杂性之间形成了显著的相互作用，产生了一款口味极其长久且享受感十足的葡萄酒，其单宁结构致密、严肃又超级光滑。贺兰晴雪、迦南美地和银色高地酒庄都推出了得分94分的赤霞珠或波尔多混酿酒，展示出了宁夏赤霞珠在成熟得恰到好处时的潜力。

品鉴中让我惊讶的是三款黑比诺和一些其他葡萄品种，包括西拉、维欧尼，还有马尔贝克。关于黑比诺，邓钟翔告诉我：在宁夏酿造出卓越的黑比诺的确是一个挑战，但并非不可能。邓是一位有抱负的酿酒师，曾在勃艮第学习，他在包括蓝赛酒庄和夏木酒庄在内的多家酒庄担任顾问。

邓钟翔说：“宁夏对于生长期较短的黑比诺来说可能太热了，尽管有这些不利因素，我还是想向世界展示这并非不可能，通过提早采收，我们不会失去黑比诺自然的酸度和微妙的特质。”他的蓝赛宁夏银川红黑皮诺2017（92分）是一个很好的例子，在偏温暖的气候框架下，也展示出了黑比诺轻盈的果香、优雅和细节。银色高地宁夏家园黑比诺2017（93分）是另一个亮点，展现了云雾般的轻盈缥缈和灌木丛的气息，尽管还有一些辛辣的新橡木特征，随后口感上酸度坚实、支撑有力，连续而宜人。贺兰晴雪宁夏加贝兰小脚丫黑比诺2017（92分）是一款更加紧凑和丰富的酒，突出了水果和颜色的深度，带有花香和轻微的葡萄梗的味道，在宽广而紧密的口感中透着清新。

再加上我们之前品尝过的香颂酒庄2017黑比诺，这些酒庄可能会颠覆你对宁夏的印象，因为他们用世界上精致但也最挑剔的葡萄之一酿造出了卓越的葡萄酒。我品鉴到的这些最好的黑比诺葡萄酒无疑打消了我对宁夏未来前景的疑虑。现在的问题不是宁夏是否有潜力生产出更优雅的葡萄，而是生产者/酿酒师是否有意愿和决心培养塑造它们；他们是否敢于走一条艰难和不寻常的道路，比如生产一款卓越的黑比诺葡萄酒。

关于宁夏的另一些亮点是葡萄品种。我们已经知道马瑟兰在中国势头强劲，最好的马瑟兰葡萄酒通常呈现出浓郁、幽深的颜色，带有柏油、辛辣的黑色浆果和蓝色花朵的特征。位于河北的中法庄园用马瑟兰酿造出了其最重要的酒款之一。在宁夏，马瑟兰也日渐流行，有些酒流露出具有异域风情的黑胡椒、薰衣草和近乎荔枝干的特点，而差一些的马瑟兰葡萄酒有点过熟，有太多果干、酒精和果酱的味道。本报告中最好的马瑟兰分别来自嘉地酒园和夏木酒庄，这两款葡萄酒都是2019年的橡木桶样品，2019是宁夏一个极佳的年份。

除了马瑟兰，西拉也在上升。容园美酒庄是由邓钟翔担任顾问的另一个酒庄，酿造出令人印象深刻的西拉和出色的马尔贝克。考虑到宁夏的高海拔和干燥的气候，让人想起门多萨，马尔贝克可能会这里流行。年轻的酿酒师梁宁用他的甘麓酒庄西拉限量版干红葡萄酒2017（93分）展示了宁夏西拉的巨大魅力。这是一瓶非常有前途的葡萄酒，融合了浓郁与平衡。有突出的蓝色水果的集中度、香料特征的橡木香和少许辛辣的黑胡椒。利思酒庄是一家有房地产背景的酒庄，是另一家其西拉值得关注的酒庄。

中国仍然是一个红酒市场，白葡萄酒并不是每个酒庄都会考虑的，但如果酒庄想要生产白葡萄酒，霞多丽是最受欢迎的品种。即便如此，夏木酒庄和美贺酒庄的几款维欧尼葡萄酒已经让人大开眼界，展示了其花香、芳香的品质和品种声望。试试夏木酒庄的宁夏维欧尼2019（93分），它充满了柠檬草、柑橘和桃子的诱人芳香，口感流畅，带着充满活力的酸度，为水果感带来更微妙并具有白垩土特征的收结。

这些葡萄酒也表明，年轻的酿酒师们正在逐渐超越他们从导师那里学到的东西，他们在宁夏赤霞珠铺陈的画布上点缀上新品种、新理念和新兴趣，正在为一个比自己年龄大不了多少的产区注入新活力。

随着葡萄酒的蓬勃发展，宁夏现在计划让酒庄更多地涉足旅游业。政府从 2013 年开始实施一个新的分级标准，酒庄接待游客的能力也被考虑在内。如果有一天你计划去宁夏旅行，有一个酒庄是不可错过的。从银川向南驱车一小时，你会看见全新的西鸽庄园，一座隐藏在戈壁荒漠中的宝石，有着令人惊叹的建筑和令人信服的葡萄酒。如今，这个拥有前卫设计以及 1 300 公顷葡萄园的雄心勃勃的现代化酒庄，目标是量产优质葡萄酒。其中，西鸽酒庄宁夏玉鸽单一园蛇龙珠葡萄酒 2018（93 分）展示了蛇龙珠在宁夏充分成熟时的巨大魅力。之前，长期以来，该品种因其强烈的草本植物、叶子味儿和病毒感染问题被认为是没有前途的品种。利思酒庄 2017 年份"義"蛇龙珠获得了 92 分，这款美味的葡萄酒表现出了品种的典型性，具有突出的风干的草本植物和收结时辛香的水果味儿。这款酒也是本报告中性价比最高的酒之一，我以人民币 199 元（约合 30 美元）的价格购买了这款酒，这对宁夏来说是非同寻常的，宁夏绝大多数优质葡萄酒的售价要比这贵很多，很多酒至少都需要 300 ～ 600 元（45 ～ 85 美元）。

廖祖宋说："蛇龙珠因为难以成熟，所以需要采收得足够晚。"廖祖宋是西鸽酒庄的一位年轻且非常靠谱的酿酒师。他以前在贝斯菲利普酒庄工作，现在与庄主张言志一起管理着西鸽的酿酒团队。张言志是一位在波尔多受过训练的酿酒师，他在奔富麦克斯的业务成功之后，就开始了这个令人振奋的项目。廖祖宋说："我们有大约 1 000 公顷超过 23 年树龄的葡萄藤。"这自然赋予他们更加浓缩果实的优势。

在宁夏，23 年的葡萄藤已经算老。留世酒庄是一个位于西夏王陵公园区的精品酒庄，也拥有数公顷 1997 年种植的这些令人梦寐以求的老藤。他们的葡萄园距离贺兰山东麓的山脚仅几英里，处在该地区的一片高地上，海拔超过 1 200 米。

对于宁夏，或者说对于中国北方的大多数葡萄园来说，另一个问题是当葡萄藤老到一定程度，埋藤将成为一项挑战。埋藤是一项花费高昂的葡萄园管理方式，但它仍然是保护葡萄植株不受低温冻害的最有效措施。即使葡萄整枝的方式已尽可能与埋土的需要相匹配，从11月初在土壤结冰之前开始的埋土防寒仍然是一项相当粗暴的做法，老藤在这个过程中更容易被折断。

甘麓酒庄的酿酒师梁宁说："当葡萄藤达到30年的时候，主干太硬、太粗，不适宜埋土，这对葡萄藤来说也是个问题。"

然而，西鸽酒庄的廖祖宋告诉我，他们提前做了计划。"我们已经为老藤预留了新梢，这些新梢将来会成为树干，这样我们在需要时就可以切断老藤进行更新，而保留原来的根系。"即便如此，对于中国北部那些需要埋土的地区来说，葡萄藤的寿命是不可预测的。话虽如此，在雪后一个寒冷的早晨，当你的目光落在葡萄园里那些孤零零的石质立柱上，那些立柱就像守卫着被土壤和大雪覆盖着的"空"葡萄园的整齐列队的士兵，这也是特别的景象，有真正的宁静与庄严之美。

宁夏劳动力的第一次增长

作为一个刚刚起步的葡萄酒产区，宁夏如果没有地方政府的帮助，不可能在如此短的时间内（10～15年）取得如此大的成就。地方政府的参与也鼓励了从能源企业和家电企业到房地产开发商等来自其他行业的投资。这让我们回到宁夏的列级庄体系，这是自2013年以来的一项官方排名，在2019年更新的名单中共有37家酒庄。"快"对中国来说已不是新闻，而当地政府希望通过鼓励酒庄酿造更好的葡萄酒以专注于长期发展，这就是为什么要建立从波尔多学来的酒庄列级系统的原因。然而，一些重要的名字却不在这个名单中。

与我交谈过的大多数酿酒师相信，对于像宁夏这样的新兴葡萄酒产区来说，列级庄排名是值得称赞的，至少利大于弊。但也存在不同的声音。我们在正在建设中的银色高地酒庄的文化中心喝着热茶，高林说"现在就急着对酒庄进行排名，还为时过早。与波尔多相比，宁夏只是个小婴儿。如果一家现在被入选列级的酒庄不了解他们的土地，市场定位不当，有一天倒闭了，那该怎么办？这肯定是件该办的好事儿，但我们不能着急。"作为宁夏领先的精品酒庄和家族企业，银色高地并不在列级酒庄的名单上，而且可能会在很长一段时间内保持这种状态。令人好奇的是，另一个非常认真的酒庄嘉地酒园也没有被列入2019年的名单。也许，正如高林所说，现在还不是时候。

高林补充说："应该有一些自省，应该听到不同的声音，或者至少我们需要冷静下来，思考一下，以免我们在考虑清楚之前就兴奋起来，每个人都毫无保留地跳进去。"

高林先生说得有道理。宁夏的葡萄酒产业刚刚离开起跑线，它将是一场马拉松。许多酿酒师也告诉我，宁夏只有少数酒庄是盈利的，大部分仍在挣扎着生存的状态。

邓钟翔却没有什么顾虑，他说他是这个列级的"全心全意的支持者"。邓钟翔大笑着说："现在有这么多的酒庄和品牌，这个列级就像一本宁夏葡萄酒指南，它肯定能帮助消费者从众多的酒庄和品牌中挑选出更好的葡萄酒。当然，这一体系有不完美之处，引起了很多批评，但这是一个良好的开端，哪一个伟大不始于批评和怀疑呢？"

对宁夏列级酒庄的评定是依据当地政府的正式文件《宁夏贺兰山东麓葡萄酒产区列级酒庄评定管理暂行办法》所制定的规则。每两年对酒庄进行一次多角度的评定，包括葡萄园（酒庄必须拥有5年以上的葡萄藤）、产量、价格、品尝小组的感官评价、获奖情况和品牌的市场价值（包括酿酒师的声誉）、葡萄酒旅游的接待能力等。每个酒庄最终都会得到一个汇总的分数，之后，可能会得到晋级或降级。新酒庄可以进入，已有的酒庄也可以被移除。邓钟翔评价道："这是一个相当动态的榜单。"

在去年的一次线上研讨会上，我记得高源说过银色高地酒庄仍然是一个家族企业，由于没有接待设施，没有被列入列级酒庄的名单。像博纳佰馥酒庄这样规模极小且注重品质的酒庄可能也会被排除在这个葡萄酒"名人堂"之外。更重要的是，这取决于各个酒庄的动机以及他们处理行政层面的荣誉和认可的意愿，这些事情不大会发生在葡萄园或酒窖，而是发生在办公室或在电脑前填写申请表的过程中。毕竟，酒庄得先申请参加才行。

在我访问中国电器巨头美的集团旗下的美贺酒庄期间，酒庄的酿酒师周兴也分享了他对酒庄列级的一些见解。

周兴说:"每一次升级的重点都略有不同。对于五级庄,关键是葡萄园,要求酒庄拥有自己的葡萄园,葡萄藤的树龄要超过五年。四级庄在葡萄栽培之外还考虑了酿酒技术。对于三级庄,你还必须包括市场、价格和接待设施。至于二级庄,品牌形象和声誉也被考虑在内。"

美贺酒庄在四级庄中总得分最高,为150.1分,似乎是升为三级庄的有力候选人。目前,三家酒庄位于金字塔的顶端(二级庄):志辉源石酒庄(180.92分)、贺兰晴雪酒庄(180.42分)和巴格斯酒庄(174.73分)。宁夏更新的列级庄名录预计将于今年很快推出,如果宁夏方面有意的话,将是宁夏有机会为其首个一级庄铺开红地毯。

试金石年份

我11月的宁夏之行绝对是收获满满、获益颇丰。每个酒庄都有很多故事,我们不可能在一份报告中全部涵盖。然而,没有什么比一瓶葡萄酒本身更能让我们回忆起一年中经历的四季了。对宁夏来说,年份之间自然会有一些起伏,但顶级酒庄一直在努力保持品质的稳定。

贺兰晴雪酒庄的张静说:"2016年不同酒庄之间的差异理应较小,葡萄的成熟度很好。但我认为,如果你真的想看看酒庄的能耐,2015年、2017年,尤其是2018年这样少见的年份是更好的试金石。"

与炎热的2017年相比,霜冻和雨水突出的2018年对于宁夏来说是一个非常具有挑战性的年份。美贺酒庄的周兴表示:"这是最凉、最潮湿的年份之一,平均降水量约为490毫米,而2017年只有大约270毫米,而极好的2019年份仅大约170毫米的降水。"在这片尘土飞扬的戈壁沙漠平原上,海拔达到1 000多米,年平均降水量通常不到200毫米,但大部分降雨都发生在夏季,灌溉是必须的。

"2019年将是宁夏的一个优秀年份,"刘建军说,他仍然靠购买葡萄来酿造葡萄酒。他的停云酒庄宁夏红胡子2017年份(93分)去年在中国酒中是亮点之一。面临2018年这样的年份,刘建军不愿妥协,他将自己所有的葡萄酒降级,以第二等级的酒标白鲸出售。贺兰晴雪酒庄的张静也告诉我,为了保证赤霞珠的品质,他们在2018年不得不舍弃近一半的红葡萄酒品种的收成。

从品酒的角度来看，在2018年取得成功的葡萄酒优雅而平衡，尽管它们缺乏一点强劲感，一些较弱的葡萄酒就是这样，中段偏水的口感被宽阔的单宁所淹没。鉴于2018年单宁的天然浓度较低，酒庄需要格外小心地进行单宁的提取。但是酿造更为细腻的葡萄酒是可能的，最好的酒庄知道什么时候应该放弃一定的浓郁度，以换取新鲜、爽脆的水果口感和比2017年份更自然的酸度，2017年份的许多红葡萄酒的特点是饱满富裕和高酒精度——大约15.5%甚至更高。

也就是说，顶级酒庄在2017和2018年份都酿造出了平衡而灵魂饱满的葡萄酒，具有健康的果实浓郁度、平衡性和单宁质量。当我们谈到气候、年份和风土，以及在葡萄园中发生的所有奇迹和努力时，这一切最终落到酿酒师如何来定义年份，以及如果有"特色风格"的话如何为酒款的特征定调。那么，我们是否应该让气候和风土完全决定葡萄酒的风格呢？可能并非如此。

值得注意的是，有些人总是按照同一个套路酿造果酱感、厚重和高辛烷值的红葡萄酒来讨好市场。他们喜欢"投机取巧"，提供给没有经验的消费者容易感受到的口味，而不是专注于精妙和细致入微的复杂度。毕竟，中国总体上还不是一个成熟的葡萄酒市场，葡萄酒鉴赏才刚刚起步。WSET项目起到了助力的作用，但在撰写本文时，它已经不得不暂停了在中国的所有课程。

对于宁夏来说，非常甜的水果味、柔顺、圆润的单宁和适中的酸度可以吸引国内市场的追随者。这并没有什么错，因为葡萄酒本质上是要卖给消费者的产品，很少酒庄有足够的时间和金钱来教育大众追求葡萄酒的复杂性、深度、精致感和结构。而且，酒庄必须做出极致的努力才能生产这样的葡萄酒。所以，这一切都归结到酒庄目前的目标消费者是谁。

"我们很多人还不清楚这一点，这对宁夏来说是个问题，"迦南美地酒庄的庄主王方去年在接受采访时说。随着葡萄酒质量越来越好，现在似乎是时候让酒庄更多地了解市场和饮用他们葡萄酒的消费者了。

宁夏在未来的5～10年里，将真正有机会在世界第二大葡萄酒市场蓬勃兴旺，这是国际葡萄酒及烈酒研究所和波尔多国际葡萄酒及烈酒展览会的预测。现在要由宁夏的酒庄来决定，是想推出许多人认为主流消费者喜欢的果味浓郁、极度肥硕的葡萄酒风格，赚些快钱，还是想以一种能长期吸引消费者的方式，生产出有新鲜感、深度、细致入微的复杂度和真实结构的更好的葡萄酒，以合理的价格销售。这对规模较小、资源较少的酒庄来说，可能是一条更具挑战的道路，但随着中国葡萄酒日见成熟和宁夏在产业的中心地位，这种想法上的不同将是顶级酒庄和优秀酒庄的分野。

2015

Meininger's Wine Business International
Inside Ningxia

wine-business-international.com/wine/general/inside-ningxia

By Jim Boyce
September–October 2015

The Chinese government is backing efforts to expand the wine sector, explains Jim Boyce, seeing it as a partner in holding back the desert. This has allowed Ningxia to develop rapidly and dynamically.

The wine regions of Ningxia appear as a green flash in a brown swath on a relief map of northwest China. The best-known ones nestle between the Helan Mountains, which blocks harsh winds from the west, and the Yellow River, which provides crucial water from the east.

Even five years ago, most people would reply "huh?" if you mentioned Ningxia. Now a steady flow of winemakers, viticulturalists, consultants, and investors pour into the region. Wine grape coverage grew from just under 3,000 ha in 2005 to nearly 40,000 ha a decade later, according to a 2015 report by Hao Linhai, Li Xueming and Cao Kailong, the three officials most associated with the region's wine sector. Given the government's goal of reaching 66,000 ha by 2020, there is plenty to do. Cao says the number of established wineries now stands at 72, with 58 more in the queue, and there are plans to add more boutique operations.

A region emerges

Some 30 years ago, Ningxia's wine industry provided bulk to producers like Changyu and Great Wall in the eastern provinces of Shandong and Hebei. Then came the 1990s, when the government became interested both in reclaiming barren land and in making value-added local brands. Efforts were boosted in 2001 when former government official Rong Jian founded the Ningxia Grape Industry Association. But the biggest change came in 2012, in line with the government's twelfth Five-Year Plan, which aims to further develop wine regions, particularly in northwest China, a mission with the added benefit of confronting desertification. As the World Bank wrote in 2012 when approving a $80m loan to help fight encroaching sand in Ningxia, desertification affects nearly 3m ha, or 57%, of the region.

In 2012, Cao Kailong established the Bureau of Grape and Floriculture Industry, since renamed the Bureau of Grape Industry Development, to help drive change. The results came thick and fast, because Ningxia offered both good grape-growing conditions and better general labour and land costs than those of eastern wine-producing provinces. Investors attracted by the ability to lease large tracts of land and grow their own grapes, rather than having to rely on buying fruit of varying quality from farmers, moved into the region.

Ningxia also made great strides in connecting to the wider wine world: it became the first region or province in China to become an OIV observer (Yantai in coastal Shandong province had joined at a municipal level) and sent delegates to annual OIV meetings. Fact-finding groups checked best wine practices in countries such as New Zealand, Australia, France and the US. Ningxia also hosted wine conferences, festivals and trade fairs in Yinchuan, including SiteVinitech—the international trade fair for fruit growing, including wine grapes—and facilitated the import of vines and the establishment of a nursery. Other projects included creating a winery classification system and organising a series of two-year projects called Ningxia Winemakers Challenge. The current challenge, with $112,000 in cash prizes, pairs local producers with 48 international winemakers to facilitate cultural and viticultural exchanges.

It would be safe to say few wine regions have seen so much activity in so little time. Arguably symbolising these efforts is the monumental Ningxia International Wine Trade and Expo Centre that opened last year. Federation president Hao Linhai said the centre was for "promoting cooperation with members of the OIV, providing a permanent wine museum, and popularising the wine of Helan Mountain's East Foothill."

Wine quality

Select Ningxia wines have won medals in local and international contests and received good reviews from critics such as Michel Bettane, Thierry Desseauve, Jeremy Oliver, Andrew Jefford and Karen MacNeil. When Jancis Robinson MW tasted 40 local wines in Ningxia in 2012, she rated five as excellent and only six as commercially unacceptable, with the main fault being oxidisation-something she noted is relatively easy to fix. A wine from Helan Qing Xue also made global headlines when it won a Decanter International Trophy in 2011. Although a common criticism is that Ningxia lacks diversity, with a default

Bordeaux-inspired Cabernet-driven style, the region has at least established it can make good wine.

Ningxia's wine region is also special because it's compact, with key wineries sprinkled along the base of the Helan Mountain range, within an hour drive of the capital, Yinchuan. Compare this to Shanxi province, where Grace Vineyard, arguably the country's best winery, stands virtually alone. Or to vast Xinjiang in the distant northwest, where wineries are often hours apart.

"The closeness of the wineries allows for a lot of communication, whether on the winemaker, viticulturalist or management level, and this is important for a developing industry," says Ma Huiqin, a professor at China Agricultural University in Beijing, who accepted an official role with the bureau this year. Ma also cites "tourism efficiency" as a benefit. "People can visit three unique wineries in a half day, from a sparkling wine producer like Chandon to a place like Kanaan, which makes more German-style wines," she says.

Ningxia's wineries encompass a diverse range of business models. These include the local branches of massive wine producers: COFCO, maker of Great Wall, has a 1,500-ha operation called Yunmo, while Changyu has teamed with Austrian winemaker Lenz Moser on a Loire Valley-esque chateau and winery, complete with museum. Foreign-funded projects are led by drinks heavyweights like Pernod Ricard, with its Helan Mountain brand, and Moët Hennessy with its Chandon operation—which produced its first commercial sparkling wines last vintage—and lesser-known entities like Delong, by Daysun Investment in Thailand, with over 6,000 ha of vineyards.

There are operations, including Silver Heights, which anchored Ningxia on the world wine map with its first vintage in 2007, and Helan Qing Xue, the aforementioned Decanter award winner. Add dozens of other wineries—from veterans Xi Xia King and Guangxia to ice wine-producing Sen Miao—and no other region in China can touch Ningxia for its diversity.

This diversity will only grow, given a focus on establishing more small operations, particularly those with production of 150,000 to 200,000 bottles, to explore the region's potential.

"It's like calligraphy," said Cao in a China Daily interview. "Everyone has his own style."

The challenges

Among the biggest challenges facing the region might be its quick success. Expectations are now higher and critics less forgiving than five years ago when the region first popped up. Critics may concede Ningxia can make good wines but question whether it can achieve greatness. French writers Bettane and Desseauve say the predominance of sandy matter and lack of clay, and especially deep soil, will result in supple fruity wines but perhaps not complex ones. And there is a not-altogether friendly green streak found in some of the wines, particularly in Cabernet Gernischt, the local name for Carménère. This may, however, be due to viticultural practices, in which case it's a problem that can be solved.

Li Demei, a professor at Beijing University of Agriculture and chief consultant at Helan Qing Xue, has long been saying Ningxia faces a long trek before it can approach the likes of Bordeaux.

How long? His protege, winemaker Zhang Jing, measures that trip in decades. "We need to make wine without defects, not just one winery, but the whole region, for 30 to 50 years, and we can slowly refine something from terroir and say this is a Ningxia wine," she says.

Zhang responds to comments about the sandy terrain by noting there is a far wider range of soil to be explored. Putting this into practice is Silver Heights, arguably China's first "garage" winery. "After many years of making wine in Ningxia, my father and I picked a different area for our grapes," says winemaker Emma Gao about the family's new vineyards, set in a much more isolated and rockier area of the region.

One thing that has to be lived with, rather than solved, is the climate. Winters are not only cold and dry, but the region suffers sandstorms and dust storms as well, and vines must be buried in the autumn and uncovered in the spring. While the burying saves the vines in the short term, it comes at the price of killing some while also raising costs. Water scarcity is also an issue, particularly as the industry grows and competes with other crops; this year, the Ningxia government announced the start of a $76m drip irrigation project using Israeli technology, which will irrigate 9,200 ha of land from 80 wineries.

Labour is scarce, too, as young workers gravitate toward urban areas. Farmers have been relocated by the government from poorer mountainous areas of Ningxia to wine-producing ones, to help alleviate the labor shortage, states that 2015 report by officials Hao, Li and Cao. "[The farmers] are trained as wine grape growers, [and] the family living standard and income are significantly improved," they write, although the sustainability of that effort, and prospects for finding further labor, remain to be seen. Not surprisingly, newer operations are more likely to use mechanised viticulture.

And while it was once thought Ningxia would remain disease-free due to its dry climate, better identification has revealed more problems with leaf roll virus and trunk disease than were previously thought. This became evident in 2012 when unusually wet weather led to widespread mildew and forced a deeper look at vineyard disease.

Finally, the prices of the wines tend to be prohibitively high by local standards—the bottles tasted by critics are often priced at $50.00 or more. It's one thing to make reasonably good wines that international critics will rate as fault-free, it's another to sell them at a price point that can compete with the influx of good, inexpensive, imported options. Such prices might work if Ningxia can establish itself as a kind of Napa Valley of China. Given the focus on quality by the local authorities, including Hao Linhai, who has the added influence of a resume that includes stints as vice chairman of the People's Government of Ningxia and as mayor of the capital Yinchuan, there is little doubt this is the goal.

Despite all the challenges, hopes are high and the learning curve steep. Practices such as picking grapes too early or using new French oak by default have quickly shifted. "A few years ago, we were worried about getting enough sugar, but now that's easy," says winemaker Zhang. "Our focus now is acidity and fruit maturity." Some producers are also moving past Bordeaux blends, with more Rieslings, Pinot Noirs, Syrahs, Marselans and even a Pinot Blanc popping up. And there is research on which grapes work best, including varieties that do not require burial, as well as on everything from irrigation techniques to canopy management. But perhaps most inspiring about Ningxia is the cooperation between the many winemakers in the region. They regularly meet to taste each other's creations and talk about how to make the region's wines better as a whole, even as each contemplates how to win its next gold medal.

The New York Times
China's Winemakers Seek Their Own Napa Valley

nytimes.com/2015/11/08/business/international/chinas-winemakers-seek-to-grow-their-own-napa-valley.html

By Jane Sasseen
7 November 2015

Taking a cue from that boutique-winery model, Ningxia has ambitions to become the Napa Valley of China. Local winemakers have won prestigious awards, and plans are underway to double the region's vineyards and create a wine tourism hub. Foreign investors have also taken notice. The French Champagne maker Möet & Chandon makes sparkling wines there, while the spirits giant Pernod Ricard is spending heavily to modernize its local winery.

"People know Napa makes the best wines in America and Bordeaux makes the best wines in France," says Hao Linhai, a top regional official who oversees the industry. "When they think of Chinese wines, we want them to think of Ningxia."

"They've got all the money in the world, they've got all the ambition in the world, and they've hired all the top consultants," says Steven Spurrier, the British wine merchant who organized the "Judgment of Paris," the 1976 blind tasting that stunned the wine world when California wines beat the French. "It's inevitable the Chinese are going to make better and better wines." (This is an excerpt. Use the url at the start for the full article.)

Hawke's Bay Today
Bay Winemaker Aims to Make His Mark in China

nzherald.co.nz/hawkes-bay-today/news/bay-winemaker-aims-to-make-his-mark-in-china/SQRNHJ3G4PTH6GGC536QMIKOH4/

By Amy Shanks
4 September 2015

It's a long way from Te Awanga to Ningxia in Northwest China but when there is wine to be made distance and location is not really an issue.

And for Clearview Estate Winery's newest winemaker Matt Kirby it could be a nice little feather in the cap should what he creates during his China wine experience get the thumbs up from a judging panel in two years—not to mention a nice little earner.

Mr Kirby, with 59 other winemakers from around the world, will be competing for the more than $200,000 prize pool in the second Ningxia Wine Challenge, which has been organised by the International Federation of Vine and Wine of Helan Mountain's East Foothill with support from Ningxia's Bureau of Grape Industry Development.

The 33-year-old, who joined Clearview at the start of the year, is one of only seven Kiwi winemakers accepted from more than 140 applicants around the world, and is the only one from Hawke's Bay.

Their tasks will be to create a winning wine made from the grapes of China's most promising wine region.

Mr Kirby will fly out for Ningxia, which is 1,100 km west of Beijing, on September 18 and will spend 15 days there taking part in grape selection and initial fermentation.

He will return up to six times over the two-year period of the challenge for between five and 15 days, to manage his wines until the autumn of 2017 when they will be judged.

All his visit costs are covered by the organisers.

"I'm really excited by this challenge," he said.

"China's wine industry is developing so quickly, it's going to be interesting to see and experience this first-hand—and because it is all so new all the technology will be the very latest, which is really exciting."

As well as the challenge of creating wines from one of China's fastest growing regions, Mr Kirby said building relationships was also behind his decision to go for a place among the global team of winemakers.

He said Clearview had been exporting to China for eight years and strong distribution networks were in place, and through that time it had become clear how much value was placed on strong relationships.

"I am sure that telling the Clearview story, along with Hawke's Bay and New Zealand wine stories, will be part of my work there."

Wines from the Ningxia region have earned high praise in China, which has a burgeoning wine market, as well as around the world.

Decanter
DAWA 2015: Ningxia Chinese Wines Scoop Gold Medals

decanter.com/wine-news/dawa-2015-ningxia-chinese-wines-scoop-gold-medals-273299/

By Chris Mercer
3 September 2015

Ningxia Cabernet Sauvignon has won two gold medals so far at the Decanter Asia Wine Awards 2015 judging week, emphasising the growing stature of the Chinese wine region.

Few people outside of China could claim to have heard of Ningxia wine 10 years ago, let alone Ningxia Cabernet Sauvignon, but investment in vineyards and winemaking mean the Chinese region has spearheaded a new wave of quality producers. (This is an excerpt. Use the url at the start for the full article.)

Wine-Searcher
Solving the Chinese Wine Puzzle

wine-searcher.com/m/2015/02/solving-the-chinese-wine-puzzle

By Claire Adamson
18 February 2015

Ningxia is one of the best spots. The province's vineyards are along the banks of the Yellow River in the shadow of Helan Mountain, where the climate is terrifyingly continental but has been tamed into submission. The high altitude and the sandy, alluvial soils count for much, but China's sheer people power is probably the most important part of the terroir. Yinchuan, a veritable hamlet with a small-town population of two million, provides enough labor to ensure that the vines are cosseted in the freezing continental winters—each vine is buried in the fall for insulation and then uncovered in spring. (This is an excerpt. Use the url at the start for the full article.)

年份的足迹 II：国际媒体报道宁夏葡萄酒

The Wall Street Journal
Move Over, France—Here Comes China

wsj.com/articles/BL-CJB-26768

By Wei Gu
7 May 2015

Quality is improving and Chinese wines have started to win awards outside China. Three wine experts in a blind tasting at a Wall Street Journal subscriber event said they were pleasantly surprised by a Bordeaux-style red from Silver Heights, a winery in Ningxia province, one of China's most respected wine regions. (This is an excerpt. Use the url at the start for the full article.)

Gourmetwelten
Mathias Regner Big in China

nikos-weinwelten.de/beitrag/mathias_regner_big_in_china_ningxia_winemakers_challenge_2015/

By Von Katharina Haase
8 November 2015

Ni hao and gan bei—a young Austrian participates in the Ningxia Winemakers Challenge 2015.

At 22, Mathias Regner is the youngest participant and the only German-speaking participant in the "Ningxia Winemakers Challenge 2015", NWC for short. Most of its competitors are between 30 and 50 years old. In Yinchuan, the capital of the Chinese province of Ningxia, he has just harvested and processed "his" Cabernet Sauvignon. There is a total of 840,000 yuan (around 115,000 euros) in prize money! (This is an excerpt. Use the url at the start for the full article.)

Indian Wine Academy
Two Indians at Ningxia Wine Challenge

indianwineacademy.com/item_3_675.aspx

By Subhash Arora
8 December 2015

An unprecedented viticultural and cultural experiment is underway in the Ningxia region of China where 48 winemakers from 18 nations including two from India—Parikshit Pramod Teldhune and Priyanka Kulkarni—are practicing their craft as part of a two-year contest of winemaking with over US$110,000 in cash prizes. (This is an excerpt. Use the url at the start for the full article.)

2016

CBS News
China Makes Big Bet on Turning Desert into Wine Region

cbsnews.com/news/china-aims-become-top-wine-producer-ningxia-region-vineyards/

By Seth Doane
1 January 2016

"I've been to every other wine region in the world, and I thought, wine near the Gobi Desert, impossible, right? Unthinkable. But, boy, wine near the Gobi Desert—it is a reality, and it's a big reality," wine expert and author Karen MacNeil said.

Now, MacNeil is updating her book "The Wine Bible," writing for magazines and trying to understand these really "new world" wines. (This is an excerpt. Use the url at the start for the full article.)

The Telegraph
Red Dawn for Chinese Wine

telegraph.co.uk/food-and-drink/wine/victoria-moore-red-dawn-for-chinese-wine/

By Victoria Moore
17 September 2016

Andrew Shaw, group wine buying director at Conviviality, signed up Changyu-Moser when he travelled to China in spring and says he intends to buy wines from other estates, too....

"Tasting, I feel Chinese wine has probably doubled in quality in two years—and the rate of improvement is accelerating. There's a huge potential blind spot on the winemaking planet that either has arrived or is going to arrive, and whoever gets in first will be able to own it. We can't afford not to be in there." (This is an excerpt. Use the url at the start for the full article.)

Decanter
China Grows Wine in Space to Beat Harsh Climate

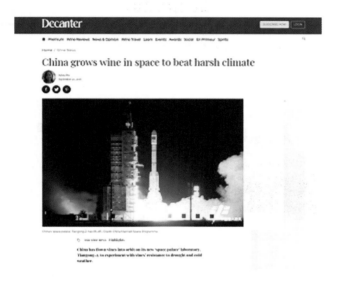

decanter.com/wine-news/china-grows-wine-space-beat-harsh-climate-331421/

By Sylvia Wu
20 September 2016

China has flown vines into orbit on its new 'space palace' laboratory, Tiangong-2, to experiment with vines' resistance to drought and cold weather....

The vines came from a nursery based in Ningxia's Helan Mountain East region, one of China's most renowned quality wine regions, reported Ningxia local media.

The nursery is owned by the Chenggong Group, which has been importing vines from France's Mercier Group since 2013. (This is an excerpt. Use the url at the start for the full article.)

Le Monde
En Chine, l'or Rouge du Ningxia

lemonde.fr/vins/article/2016/10/24/en-chine-l-or-rouge-du-ningxia_5018983_3527806.html

By Simon Leplatre
24 October 2016

Ces dernières années, quand un vin chinois gagnait un concours, il venait la plupart du temps du Ningxia. Cette petite région autonome, coincée entre le Gansu et la Mongolie intérieure, au centre du pays, semble se spécialiser dans les vins de qualité. (This is an excerpt. Use the url at the start for the full article.)

Hermanus Times
Local Winemaker Impresses Chinese Connoisseurs

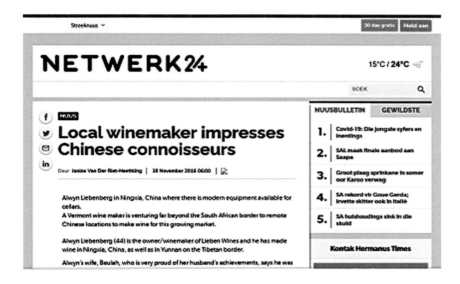

netwerk24.com/ZA/Hermanus-Times/Nuus/local-winemaker-impresses-chinese-connoisseurs-20161123-2

By Janine Van Der Riet-Neethling
28 November 2016

A Vermont wine maker is venturing far beyond the South African border to remote Chinese locations to make wine for this growing market.

Alwyn Liebenberg (44) is the owner/winemaker of Lieben Wines and he has made wine in Ningxia, China, as well as in Yunnan on the Tibetan border.

Alwyn's wife, Beulah, who is very proud of her husband's achievements, says he was chosen as a finalist in the Ningxia Winemakers Challenge in China last year, to make wine for a competition that runs over two years and will be judged in September 2017 by an international panel...

Beulah says Alwyn is extremely happy with the wine he produced, as it has the Bordeaux character he wanted. "It gets so cold in Ningxia that the entire vineyard is buried under sand during winter so that the vines do not freeze and dug up again when it gets warmer. They even have heating in the cellars to protect wine from freezing in winter." (This is an excerpt. Use the url at the start for the full article.)

2017

Marlborough Express
Mayor Takes Trip to China to Sign Sister-city Agreement

i.stuff.co.nz/business/96114564/mayor-takes-trip-to-china-to-sign-sistercity-agreement

By Jennifer Eder
30 August 2017

New Zealand's largest winemaking region is about to cement a sister-city agreement with Chinese winemaking region Ningxia.

Marlborough Mayor John Leggett and representatives from Marlborough's Sister City Committee will finalise the deal in a visit next week...

Marlborough Mayor John Leggett said he saw similarities between where Ningxia was now and where Marlborough was 30 years ago, when grapes were a relatively new crop.

"They recognise that Marlborough is now one of the world's leading wine regions and that a great deal of knowledge and expertise resides here."

The sister-city agreement comes after several exchange visits between the regions over the last 18 months, intended to create education and wine industry opportunities for the regions.

So far, the visits have led to a Ningxia winemaker spending a vintage in Marlborough, a group of Ningxia students visiting Marlborough Girls' College and Marlborough Boys' College, and six more enrolling in a viticulture and winemaking degree at Nelson Marlborough Institute of Technology.

Some Marlborough wine technology businesses have secured contracts with Chinese customers.

Former Marlborough Mayor Alistair Sowman signed a memorandum of understanding with Ningxia last year.

Leggett said the next step was to formalise the relationship with a region-to-region agreement.

"We've been through the preliminary steps and now we have an understanding of what each region hopes to achieve from this relationship," he said.

The sister-region relationship was a great opportunity to expand the education and training delivered through Marlborough's secondary schools and tertiary institute, he said.

Committee members Alistair Sowman, Lily Stuart and Cathie Bell would accompany Leggett on the trip, which would run from Sunday to September 10.

They would be joined by winemakers Richard O'Donnell and Dave Tyney, who worked as a winemaking consultant in Ningxia for part of the year, and Ningxia-based education agent Kiki Chenshu. (This is an excerpt. Use the url at the start for the full article.)

The Buyer
On the Road: How China is Stepping up the Quality of Its Wines

the-buyer.net/insight/ningxia-china-looks-to-present-quality-wines-to-the-world/

By Victor Smart
2 October 2017

Nobody could accuse the Chinese of being unambitious. Since they have made their taste for wine felt across the globe the Chinese have started a major programme of cultivating vineyards in their own country, re-orientating middle class drinking habits and putting in place systems that they hope will ensure the wine they produce is not only best in class but is universally acclaimed as such. Victor Smart travelled to Ningxia to see first hand the scale of Chinese ambition.

The Chinese wine industry is already the stuff of legend with Ningxia winery Ho-Lan Soul turning boulder-strewn wasteland into 30,000 hectares of vines... oh, and building a wine theme park at the same time.

In many ways Ningxia, an economically poor region separated by a narrow mountain range from the Gobi desert, is an unprepossessing place. Biting winds force winegrowers to bury the vines in the winter, meaning they have to be uprooted every 20 years. And yet this is the up-and-coming region for winegrowing in China—and the pursuit for excellence is undeniable.

The top wineries won't rest until they can match esteemed wines from France on quality. A few already can.

What impresses visitors immediately is the ambition, impatience and scale. With local winemakers already making many good wines and a few excellent ones, officials are clear that the next objective is 'to take Ningxia wines global'. The local government is throwing money at promotion, including joint television series productions with Australia with storylines featuring, yes... dating winemakers! And the government, which generally wants a shift to a consumption economy, is determined to turn China's middle classes into a nation of wine drinkers.

What all this means for wine producers around the rest of the world in the next decade can only be guessed at. But winemakers from all four corners of the globe now flock to the annual Ningxia Wine Expo held each year in the region's capital of Yinchuan. There are Swedes and Brits, and Chinese translators are hurriedly being schooled in French. The view, as Austrian winemaker Lenz Moser has put it, is that 'China is too big to fail'.

At the elegantly walled Château Yuanshi we are introduced to the owner, Ms Yuan Yuan, aged 25, and her winemaker, Mr Yang Weiming, aged 31, who studied winemaking at Suze La Rousse in France. This pair, who oversee 130 hectares, could be seen as the new face of wine in China.

Tending the vines is highly labour intensive—and the need to bury and unbury the wines makes it even more so. But everything about the place shows a strong sense of flair and self-confidence.

The young wine producers make a very attractive Chardonnay, a little bit fat in the mouth, which lights up the faces of the French wine experts I am travelling with. And the château is also being kitted out expensively with traditional Chinese antiques and halls to fulfil its secondary function as a tourist attraction.

To appreciate the true scale of China's ambition, however, a visit is necessary a little further north to the Ho-Lan Soul winery. A decade ago the chairman, Mr Chen Qi, defied the advice of just about every sane person and bought thousands of hectares of desolate boulder-strewn land close to the mountain.

The vineyard now has a staggering 30,000 hectares under cultivation and is one of the biggest producers in China. What has become of the notion of terroir, I don't know, but Ho-Lan Soul's logo includes a fearsome traditional warrior which captures the brand's conquering ambitions.

Cavernous warehouses hold hundreds upon hundreds of new French oak barrels full of various vintages; the winery's Organic Shiraz 2014 performed particularly well in our blind tasting with a classic French style and the Shiraz to the fore. There is also a fine Cabernet Sauvignon 2012. Prices are high, a bottle of the latter costing around £200. It is evident that the Chinese want to be 'reassuringly expensive'.

Needless to say Mr Chen's ambitions do not stop there.

The next phase of the project will be to turn what was so recently a wasteland into an attraction on a new wine route bringing much-needed tourist revenues into Ningxia, a Hui autonomous region. The winery will soon develop into a wine theme park with private châteaux dotted about and the label-conscious super-rich able to buy, and possibly cultivate, their own plots of vines.

China is already proving a lucrative market for foreign winemakers, bottling machinery manufacturers and cork suppliers (screwcaps confer too little prestige to an expensive bottle to be considered).

One official at the wine expo happily predicted that China, with its growing middle class, will soon be the world's largest consumer of wine. With a population more than four times that of the US, which currently tops the list, that will be difficult but far from impossible. It will require the Chinese middle classes to fundamentally re-orientate their drinking preferences away from traditional products towards wine, but that's a process already well underway.

The good news for winemakers is that prestigious wines are accorded a hefty price premium. And certainly there seems to be no talk about drinkers' 'units'.

Ningxia has embarked upon a classification modelled on Bordeaux's world-renowned system dating back to 1855 ranking wines from first to fifth growths (crus). This is telling, more than anything revealing both China's determination and haste to compete with the world's most sought-after wines.

To Europeans it may seem premature that an industry little more than a decade old should want to dignify its best wines with such categories given the challenges such an invidious task implies. But one suspects in Chinese eyes the wine world is not just about producing the best but about being universally acclaimed as producing best.

Decanter
Going Biodynamic in Ningxia

decanterchina.com/en/columns/anson-on-thursday/going-biodynamic-ningxia

By Jane Anson
Translated by Sylvia Wu
26 October 2017

Jane Anson discovered a different side of China's fledgling fine wine region during a visit this month.

It's the last stop we make, in a snatched half hour before we leave for Yinchuan Airport. Not on the official programme but timings made easier because the unmarked property backs directly onto one of the main traffic arteries leaving the city.

The week so far has involved an array of estates in the Ningxia Hui autonomous region, around 1,100 km from Beijing in the sparsely-populated northwest and considered to be the fastest-growing and most exciting wine area in China.

We have seen large government-backed properties like COFCO's Great Wall, international flagships like LVMH's Chandon and Pernod Ricard's Helan Mountain and renowned boutique wineries like Emma Gao's Silver Heights and Zhang Jing's Chateau Helan Qingxue.

Most are laid out along the newly designated Wine Road that snakes along the eastern foothills of the Helan Mountains at around 1,200 m altitude.

But this is something else entirely. A tiny, 6 ha farm—but with only 2.8 ha of vines—planted by husband and wife team Peng Shuai and Sun Miao who returned to her home town of Yinchuan in 2013 after six years studying in Beaune and working with, among others, Emmanuel Giboulet on the Côte d'Or and Domaine de la Solitude in Châteauneuf du Pape.

The pair met at high school in Shandong Province, where Sun moved as a teenager, then attended the local university together before signing up for an exchange programme in France, and staying on as they fell more deeply in love with wine.

"We were offered a piece of land nearer to the mountains, but there was no electricity out there at the time and we couldn't afford to install it," says Peng.

"We chose instead this plot that was closer to the city and so more practical. We also liked the lines of trees that surround us here, and felt we could encourage more biodiversity to fit our belief in biodynamics."

It stands in stark contrast to the vast investments made elsewhere. The Pernod Ricard land, for example, was part of the Gobi desert when it was first earmarked in the 1990s and over one hundred hectares of sand dunes were bulldozed before land reclamation and irrigation channels made vines possible.

Many properties are vast reimaginings of Loire or Bordeaux, from the £22 million marble and stone Chateau Mihope that opened last year to the stunning

£70 million Chateau Changyu Moser XV with its 800-strong barrel cellar and museum to Chinese wine history.

The spending at this final stop is minimal. The winery is tiny, with four small-sized stainless steel tanks and an open-topped wooden barrel that is used as a dynamiser for turning flowers and herbs into biodynamic preparations.

The one-roomed office lies at the end of a dirt road with vines on either side-looking a little sparse in places, because only around 30% of the vines are expected to survive against disease and the freezing winters, despite being buried to protect against the cold like all the vineyards in the region.

"We buy local Ningxia vine stock that cost a fraction of imported French vines," Sun tells me, speaking French rather than English, "because right now we are simply focusing on the soils. We don't want the pressure of big investments and marketing budgets."

This is one of the few times I've heard serious talk of working on soil structure since I've arrived in Ningxia.

Vineyard management as a rule is still in its infancy, although there are signs things are changing. Chateau Changyu Moser XV started paying for grapes by quality instead of weight a few years ago, with the arrival of respected Austrian winemaker Lenz Moser.

And viticulturalist Richard Smart was brought in to Chateau Mihope to work on the planting. The estate uses new training systems specifically developed in the region to encourage full ripeness and protect vines during the winter burials. But most of the focus for investments to date has been on the often gleamingly-modern wineries.

At this tiny estate, the vines date back to 2010, planted when on a trip back from France.

And although there is far more clay here than up in the astonishingly poor sandy-gravel soils on the foothills of the mountains, it still has extremely low levels of organic matter and the soil pH is a high 9.

To regulate and bring more balance, Nicolas Joly's daughter Virginie has been sending the biodynamic treatment Preparation 500 to them, while other ingredients are secured by Giboulet in Beaune.

They have just found an organic cattle farm in Ningxia where they can buy cow horns for burial, and a fresh delivery of manure from this farm was being applied as we walked through the vines.

Given the focus on soils, microbes and what happens underground, it seems entirely appropriate that the one clear area of investment has been the digging out of an underground cellar ('we finished it in one month, Chinese-time', says Peng).

The entrance looks like a concrete bunker but walk down the sloped concrete path and you're assailed by the damp earth, mushroom cool of a French wine cellar. Eyes take time to adjust and then two rows of barrels appear, with bottles stacked neatly to one end, all labeled by vintage and style. A long tasting table is surrounded by customised barrels turned into seats.

The first time you see the name of the estate, Domaine des Arômes, is with a hand-written sign taped to the smoked-glass table top. There is no temperature control but even in a climate that can reach -25°C in winter and 40°C in summer the cellar stays between 8-18°C year round.

I've been impressed by many of the wines tasted this week, but few have made me feel so simply happy to be discovering them.

We taste just two reds from the vintage 2013, along with a red and white from 2014. One is labelled as a natural wine with no added sulphur. Cabernet and Merlot blends, as are so many in the region, with Chardonnay for the white. These are still early days, and the wines will take time to grow in stature, as more life is teased out of the soils. But there is heart here, and a real future.

The Beijinger
Annual Grape Wall Challenge Gives You A Perfect Starter List for Chinese Wines to Try

thebeijinger.com/blog/2017/02/15/annual-grape-wall-challenge-gives-perfect-great-hit-list-chinese-wines

By Jim Boyce
15 February 2017

Cui Yunan, who does marketing in the food sector, said the Chinese wine tasting showed each wine's uniqueness....

She was surprised at wines from her home region of Ningxia. "I know we have great weather for grapes, and I know we can make good wine, but I didn't know we can make it that good!" she said. "It was beyond my expectations."(This is an excerpt. Use the url at the start for the full article.)

China Daily Asia
Winemakers Rise to Ningxia Challenge

chinadailyasia.com/asia-weekly/article-13555.html

By Wang Hao & Li Xiang
16 October 2017

From the four corners of the Earth they came, armed with the knowledge, wisdom and patience that would help them and their hosts rise to the challenge placed before them. For the 48 winemakers from 16 countries, this remote area of the Ningxia Hui autonomous region, in Northwest China, became not only their second home, but also a test lab over the past two years as they worked to find the perfect drop.

　... the biggest winner of all was the wine industry of Ningxia, centered on the eastern foothills of the Helan Mountains, whose winemakers have been the beneficiaries of barrel loads of indispensable expertise and tips over the past two years.(This is an excerpt. Use the url at the start for the full article.)

2018

WineLand
Move Over Napa, Here Comes Ningxia

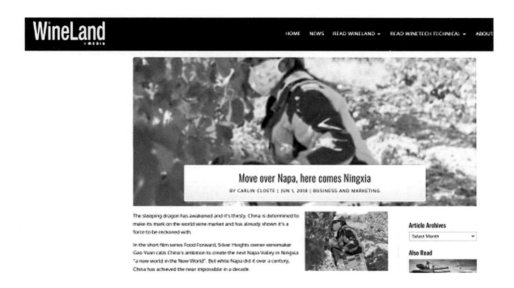

wineland.co.za/move-over-napa-here-comes-ningxia/

By Carlin Cloete
1 June 2018

The sleeping dragon has awakened and it's thirsty. China is determined to make its mark on the world wine market and has already shown it's a force to be reckoned with.

In the short film series Food Forward, Silver Heights owner-winemaker Gao Yuan calls China's ambition to create the next Napa Valley in Ningxia "a new world in the New World". But while Napa did it over a century, China has achieved the near impossible in a decade.

In 2015 Reuters reported that China had overtaken France as having the second largest area planted with vines [including for table grapes]. And what a meteoric rise it has been. Since the beginning of this century, China has doubled the land devoted to vines to just under 800,000 hectares, according to the International Organisation of Vine and Wine (OIV). In 2015 China accounted for 10.6% of the world's vine-growing area, compared with France at 10.5% and the world's number-one grower, Spain, at 13.5%.

"It's all government driven," says Lieben Wines' Alwyn Liebenberg, who consults in China as a so-called flying winemaker. He will return to the country for the ninth time later this year. "The government has decided for example that Ningxia, formerly a fairly poor coal mining area, will be the world's next Bordeaux. Ningxia borders the Gobi Desert and since most of China's population lives on the coast, they want to make this area attractive for development in order to utilise the whole country. To get prospective wine producers started, the government often gives them soft loans of which only a part is repayable. All the big companies are here: Moët Hennessy, Pernod Ricard and many of the biggest Australian labels."

The soil in Ningxia is ideal for viticulture and, despite being an arid semidesert, the Yellow River runs through the area. In some parts, drilling 200 metres deep has resulted in an abundance of fresh water. With government aid, dams are built and complex irrigation schemes installed, transforming barren landscapes into rolling vineyards. The government plans to see 200 wineries settled here by 2020, according to a New York Times article published in 2015.

Like Bordeaux, Ningxia lies on the 38th degree latitude. Gao Yuan, also known as Emma, learnt her winemaking skills in France and returned to the family farm with her French husband, Thierry Courtade, to head up Silver Heights Vineyard.

In a Wall Street Journal blind tasting involving influential tasters such as the first Asian wine master, Jeannie Cho Lee, the Silver Heights Cabernet Sauvignon, along with a Grace Vineyard Cabernet Sauvignon from the Shanxi region outperformed the likes of Châteauneuf-du-Pape, leaving the tasters astounded. It seems the Gobi Desert with its dry, gravelly soil can produce great wines if the vineyards are nurtured correctly.

"There are some beautiful wines that come from this area," Alwyn says. "But I think there are even better sites for winemaking in China. You won't find the quality in a R60 equivalent bottle as you would in South Africa. But spend R3000 and you'll really see some fantastic wines." He recently consulted for a French-Chinese company where he made wine on the Tibetan side of China. "I lived in the mountains among women who only speak Tibetan. They have an ancient winemaking culture where everything is done by hand, from picking to sorting, and the wine is bottled in urns. The wines fetched ridiculous prices and the whole harvest was snapped up in downtown Shanghai."

Of course, it's no use growing great grapes without growing the market along with it. Trying to establish yourself as the next Napa means convincing international wine buyers and getting them interested. And the Chinese are doing this in innovative ways. Mountain Ridge's Justin Corrans knew nothing about China when he was invited to take part in the Ningxia International Winemakers Challenge in 2015. It was a brilliant marketing ploy to get Ningxia in the spotlight as a premier winemaking region. Justin, Alwyn and Carsten Migliarina, also of SA [South Africa], were three of 48 winemakers from 17 countries who were invited to take part in the two-year-long winemaking challenge. Each contestant was paired with a local winery and the challenge was to make wine. Justin was paired with Domaine Lanxuan, and on 29 August last year a panel of international judges declared him one of five gold winners for his Cabernet Sauvignon and the overall winner.

"One of the challenges of making wine in a desert is the incredible temperature fluctuations," Justin says. "In winter it can often get as cold as -20°C and they have to bury the vines to protect them." Last year, China launched Cabernet Sauvignon, Merlot and Pinot Noir vines into space aboard the Tiangong-2 laboratory in the hope it would trigger mutations that would make the plants more suitable for the harsh climate and resistant to drought and viruses.

Temperature fluctuations is not the only problem in Ningxia. "China is now facing the same problem South Africa did about 30 years ago," Justin says. "They're struggling to get good clonal variety. And remember, they're all first-generation wine drinkers, winemakers and viticulturists, so they must consult widely to make up for a lack of experience. They're adapting very quickly and the industry is really only a decade old, so there's a problem to get good virus-free material."

Mike Veseth, editor of the blog The Wine Economist, writes that getting control over land is not easy in China. Many producers still have to work in large part with hundreds of small family-run vineyards. Some are coping with this difficult supply chain by building relationships with the producers. They rent land to their own labour force and provide training and credit in exchange for inputs.

China has since 2013 been the biggest consumer of red in the world, according to a report by France24 English. In Spiros Malandrakis' summary of key trends for the wine industry in 2018, published on just-drinks.com, he predicts China will accelerate the transition from wine as a luxury item to a mass-market seller, "a development that will, in turn, shape the global wine industry, both in terms of supply and demand as well as in branding and positioning".

Justin and Alwyn agree the Chinese are serious Francophiles. "If it sounds French, it must be good," Corrans says, hence the emphasis on French classics such as Cabernet Sauvignon. But Justin adds, in his opinion, in Ningxia especially they can do a lot better with earlier ripeners such as Merlot or Pinot Noir.

Author Li Demei, a regular contributor to Decanter China, says Marselan is another varietal with huge potential. He was the winemaker for Domaine Franco-Chinois when he made the first Marselan wine in China on the recommendation of French consultants. "I believe the potential of Marselan lies in its high productivity, flexibility to the environment, complex and beautiful flavours, rich and concentrated palate, and potential to be made into different styles."

In an article on the Chinese wine-growing climate, he says areas that are generally referred to as "wine regions" in China, are in fact administrative divisions. The one with the longest winemaking history is Shandong province, and particularly the Shandong Peninsula and its Yantai region. Wines from Shandong and Hebei provinces account for over half of wines in the Chinese wine industry, by both yield and output value. Other areas where wine is made are Beijing, Tianjin, Shanxi (where another internationally acclaimed winery, Grace Vineyard, is situated), Shaanxi, Jilin, Liaoning, Xinjiang Uyghur Autonomous Region, Ningxia Hui Autonomous Region, Gansu, Inner Mongolia Autonomous Region, and Yunnan and Sichuan provinces, the latter which borders Tibet.

Whether such a phenomenal growth curve can be sustained, remains to be seen. It's reassuring that the Chinese public is increasingly embracing wine as an everyday beverage, thereby broadening the total wine-buying public, which can only be good for the global wine business.

China Daily
Wine Down the Track

chinadaily.com.cn/a/201808/17/WS5b760 ecaa310add14f3863c5_1.html

By Li Yingxue
17 August 2018

Vino lovers can now embark on a voyage of discovery to Ningxia's vineyards while sampling the region's flavors along the way, thanks to a new luxury-train service, Li Yingxue reports.

All aboard! All aboard! The train departing for Yinchuan at 8 PM has one notable addition—an extra boutique carriage that houses a bespoke wine-tasting room and six luxury sleeping compartments.

A night on this train is something out of the ordinary, as passengers in the upscale carriage can enjoy a meal paired with wines produced in the Helan

Mountains' East Foothill Wine Region, followed by a wine facial mask before bed.

Launched on July 31, it's the first moving winery in China and the first overnight luxury-rail service between Beijing and Yinchuan, capital of the Ningxia Hui Autonomous Region.

The carriage can accommodate 14 passengers through its combination of two single ensuite bedrooms, two double bedrooms with a shared hand sink, and two four-person rooms—all featuring complimentary toiletries and towels.

The mobile-winery project is the brainchild of the Ningxia Bureau of Grape Industry Development, while the China Railway Lanzhou Group operates the service.

Two hundred journalists from around the country have been invited to experience the project over the course of August. It'll then run as a passenger service for the next three years, according to the project's coordinator, Yang Yang.

"We have a separate kitchen in the carriage, and we invite chefs from different wineries in the Helan Mountains' East Foothill Wine Region to take turns to cook for the passengers," says Yang.

Sommeliers pair wines with the dishes and pass on their wine knowledge to the passengers.

"After the train arrives in Yinchuan the next morning, the passengers can visit the vineyards in the region, pick grapes and sample different wines," says Yang.

The Helan Mountains' East Foothill Wine Region was recognized as a national geographical-indication protection-product area by the Chinese General Administration of Quality Supervision, Inspection and Quarantine in 2002.

In 2013, the region launched its first winery classification system to improve the regulation of the production of quality wines in China.

The area now has 86 wineries growing grapes over some 38,000 hectares with an annual production of about 100,000 tons of wine. The wineries have created about 120,000 job opportunities and attract more than 400,000 tourists each year.

Shao Qingsong, a winemaker at Lilan Winery, says Ningxia is an ideal location for grape growing. "We don't have much rain in Ningxia, and if we need water, we can transfer it from the Yellow River," says Shao.

Lilan Winery is located in Yuanlong village in Minning town, 30 kilometers southwest of Yinchuan, in the heart of the Helan Mountains' East Foothill Wine Region.

The water from the Yellow River is rich in organic matter that benefits the grapes, and the soil releases an energy at night that helps the fruit mature.

"We make our own organic fertilizer to use on our grapes instead of using chemical ones, which helps to lower costs while increasing the quality," says Shao.

Unlike other wine regions, the grapevines in Ningxia have to be buried in the soil in winter and dug up again the following spring to prevent them from dying in the extreme and arid conditions. Shao and his teammates are working on creating ways to tackle this problem moving forward.

"From our 107-hectare area for grape growing, we can produce about 500 tons of grapes, which can be made into about 400,000 bottles of wine," says Shao. "We are now selling our wines to France and England."

Shao believes Ningxia is capable of producing high-level wines, but it needs time to improve. "If you plant grape seeds this year, you will only get grapes that are good enough to make quality wine five years later. Our winery has only been producing wines for three years, so there is still some way to go."

Set up in 1997, Chateau Hedong does not suffer from the same problem. Their vineyards have more established vines, some of which date back more than a century.

After taking over the winery in 2010, owner Gong Jie studied business administration at Tsinghua University to learn more about how to manage his vineyards.

"I used to be in the mining business, so I had to learn about making wines from scratch," says Gong. "But life has been getting steadily better since I started to work with the local farmers to develop the soil, from the first shoots of spring to the harvest in autumn."

"The government is also supporting us by rewarding winemakers with 500,000 yuan ($72,600) for every award they win at the most influential wine competitions. So far, we have won three."

Gong is building a "wine town" alongside his winery, which he hopes will be a AAAA-level tourism site. (AAAAA is the highest national designation.) By offering classes about viniculture and setting up a comfortable hotel, Gong hopes tourists will stay longer at his vineyard and learn more about his wine.

"Each day we have to limit the number of tourists visiting our winery to around 200, because more people in the wine cellar could affect the temperature and quality of our wine," says Gong.

Most wineries in the eastern foothills region also operate as tourism sites.

The region is now working with a cruise company to build a winery on a ship, as well as a train.

The Ningxia wine region was included in The New York Times' "46 Places To Go" list in 2013, as "the local government has reclaimed desert—like expanses, irrigated them profusely, planted them with cabernet sauvignon and merlot and started a campaign to transform this rugged backwater into China's answer to Bordeaux".

So now, for the price of a train ticket, wine lovers can embark on an odyssey of discovery—sampling the tastes of Ningxia's wine country along the way—to take in the European style of Chateau Changyu Moser XV or the traditional Chinese desert architecture of Yuanshi Vineyard.

Gourmetwelten Das Genussportal
Auf der Wein-Route durch das Reich der Mitte: Beste Weine Chinas

nikos-weinwelten.de/beitrag/auf_der_wein_route_durch_das_reich_der_mitte_beste_weine_chinas/

By Niko Rechenberg
19 August 2018

Eiffelturm und Pyramiden wirken klein, wenn man zum ersten Mal vor ihnen steht. Durch die vielen Postkarten, Filme und Erzählungen hat man sie sich viel größer vorgestellt und ist enttäuscht. Sie kennen dieses Phänomen?

  Nicht so in China! Zwar hatten Kollegen von unglaublichen Ausmaßen berichtet, Stuart Pigott und China-Experte Frank Kämmer lagen mir bereits seit Jahren in den Ohren, endlich in das Reich der Mitte zu kommen. Erstaunliches ist passiert: Vor Ort ist in China alles noch riesiger, unglaublicher und aufregender!

Dabei sind wir in einer knappen Woche nur einen kleinen Teil der chinesischen Wein-Route abgefahren. Ningxia ist das chinesische Napa, dort haben die Weingüter die gleichen Ausmaße wie beispielsweise das Disney-Familiengut Silverado, wo locker mit dem SUV im Kamin eingeparkt werden kann. Viele Chateaux in Ningxia sehen aus, wie direkt von der Loire importiert- nur anstatt zwei Etagen sind es sieben und statt mit vier Türmchen sind die Schlösser mit einem Dutzend Türmen ausgestattet.

Ningxia entwickelt sich zur besten Weinregion in China: Dort haben wir 9 China-Chateaux besucht und Dutzende weitere Weine aus der Ningxia-Challenge verkostet. Die Wein-Route in Ningxia hat eine Länge von gut 150 Kilometern und wartet mit über 100 Weingütern auf 30,000 Hektar Anbaufläche auf. Am besten erreichbar ist die Region von Yinchuan aus, einer 2-Millionen-Einwohner-Stadt, rund 1,100 Kilometer westlich von Bejing und südlich der Mongolei und der Wüste Gobi gelegen.

Hier herrscht perfektes Sommer-Klima. Nur im Winter wird es wegen der extremen Tiefst-Temperaturen von minus 25 Grad hart für die Reben. Aus diesem Grund werden die Rebstöcke schräg wachsend erzogen und im Winter zum Schutz zugeschüttet und eingegraben. Oftmals werden die Flächen dann mit Wasser geflutet und die Rebfelder werden zu riesigen Eisflächen, unter denn die Rebstöcke überleben können.

Der Wein-Konzern Changyu Pioneer wurde bereits 1892 vom Diplomaten Zhang Bishi gegründet-dessen erster Weinmacher war der Österreicher August Wilhelm Baron von Babo. Changyu ist heute der größte Wein-Produzent in China und auch heute ist der wichtigste Berater vor Ort wiederum ein Österreicher-Lenz Moser, der jahrelang für Robert Mondavi arbeitete und heute Chateau Changyu Moser XV für Changyu führt.

Hinzu kommen chinaweit weitere Weingüter wie Changyu Afip, Changyu Castle, Changyu Rena und Changyu Baron Balboa. In der Changyu International Wine City bei Yantei stehen 95 Hallen für die Weinproduktion zur Verfügung.

Jede Halle gefühlt so groß wie die ehemalige Abfertigungshalle des Flughafens Tempelhof-insgesamt eine Fläche von 270,000 Quadratmetern, rund 250 Fußballfelder.

Es ist die größte Weinproduktion der Welt: Als ein Weingut bei voller Auslastung hochgerechnet, wird es jährlich 400,000 Tonnen Wein und Schnaps produzieren-das ist die Hälfte der Jahresproduktion an Wein in Deutschland (zwischen 7.5 bis 9.5 Millionen Hektoliter).

Der Großteil der Weine in China, weit über 90 Prozent, sind rot, den Riesenanteil liefert dazu Cabernet, dann kommen Merlot und Syrah. Alle Weine, die wir verkosten konnten, waren erstaunlich gelungen und sauber vinifiziert, die besten von ihnen hätten als Piraten in einer Bordeaux-oder Rhone-Verkostung für Aufsehen gesorgt.

Die einfachen sind für 10 bis 20 Dollar zu haben, die besseren kosten schnell 50 oder auch schon mal 100 Dollar, 150 oder deutlich mehr. Kein Problem für die vielen chinesischen Millionäre und auch die wachsende Mittelschicht in den Großstädten hat Lust auf gute Weine und Weinbars.

Über 90 Prozent des Weines wird angeblich in China selbst konsumiert. Dabei trinkt die große Masse der Chinesen nach wie vor Bier und Billig-Schnaps. Meine Vermutung für den Verbleib der riesigen Wein-Mengen: Künftig wird Europa mit dieser Produktion-natürlich zu deutlich niedrigeren Preisen-mit einer großen China-Wein-Welle per Zug über die Neue Seiden-Straße und parallel dazu über die maritime Silk Road (der Seeweg) überschwemmt.

100,000 Waren-Container zwischen Deutschland und China bis 2020 sind von der Deutschen Bahn angekündigt worden. Die ersten Routen der "One Belt, One Road" -Strecke mit insgesamt einem Bau-Volumen von 900 Milliarden Dollar sind fertig gestellt. Anfang Mai kam der erste Güterzug mit 44 Containern auf der neuen 9,800 Kilometer langen Direktverbindung zwischen China und Österreich in Wien an und wurde mit großem Pomp empfangen.

China hat eine Rebfläche von knapp einer 1 Million Hektar. Das bedeutet zunächst hinter Spanien Platz Zwei. Doch der Traubenanbau dient in großem

Maße auch der Produktion von Schnaps, Tafeltrauben und Rosinen. Trotzdem wird China in wenigen Jahren der größte Trauben-und Weinproduzent der Welt sein. Lag der Pro-Kopf-Verbrauch vor 10 Jahren noch bei zwei Deziliter Wein, liegt er inzwischen bei einem Liter-Tendenz steigend.

Was heißt das für die Wein-Route in Ningxia? Dort soll es in spätestens 10 Jahren 60.000 Hektar Anbaufläche mit 300 China-Chateaux geben, also eine Verdoppelung. Noch Fragen? Hinfahren, staunen und Wein trinken!

Auswahl der GOURMETWELTEN der besten Weingüter in Ningxia

Behalten Sie den Überblick, bald erreicht uns der Wein-Tsunami aus China: Auf der Wein-Route durch das Reich der Mitte

1. Chateau Changyu Moser XV—Teil des Changyu-Imperiums und mit die besten Weine des Landes. Sehr gutes Preis-Genuss-Verhältnis, das Flagship, der Cabernet aus 2015 holt locker 93 Punkte.

2. Legacy Peak—wow, das hat Potential, der Family Heritage Cabernet kommt auf 93 Punkte, der junge Winemaker von Legacy ist schon jetzt der "Justin Bieber" Chinas!

3. Silver Heights—Besitzerin Emma Gao und ihr Mann Thierry Courtade-er arbeitete früher auf Château Calon-Ségur in Bordeaux-erzeugen rund 50,000 Flaschen im Jahr. Der Family Reserve Chardonnay ist großartig, einer der besten der Region mit 92-93 Punkten, die Reserve Helan Moutnains ist ein ebenso exzellenter Cabernet.

4. Jiabeilan (Helan Qingxue Vineyard), der 2009er Cabernet Sauvignon Reserve Jiabeilan von Helan Qingxue Vineyard von der quirligen Jing Zhang (Foto unten) gewann die Decanter Wine Trophy und wurde so zum Meilenstein für den chinesischen Wein. Ausgezeichnete Qualitäten!

5. Kanaan—ein christlich (!) geprägtes Weingut, ebenfalls von einer

Weinmacherin, Fang Wang geführt und promotet: Pretty Pony, Wild Pony und das Flagship Black Beauty mit aufsteigender Qualität überzeugen und machen richtig Spaß!

6. Lan Cui—gut gemachte Weine eines Modehaus-Besitzers.

7. Dong Li ist im Kommen—quasi der Aufsteiger der Region, Flagship ist der Family Treasures mit sehr anständiger Qualität.

8. Yuanshe Stonecastle Vineyard—Die Weine heißen Son of Mountains und Soul of Mountains, nette Umschreibungen für Weine, die sich noch strecken müssen, um in Europa Erfolg zu haben.

9. Chandon—Noch nicht die große Champagner-Klasse wie bei uns, aber der Rose-Brut macht sich schon jetzt. Die Kosten für die 4 Qualitäten: um die 20-25 Dollar pro Flasche.

jancisrobinson.com
Young Guns of Ningxia

jancisrobinson.com/articles/young-guns-of-ningxia

By Louise Hurren
2 October 2018

Louise Hurren sheds light on the rapid evolution of the Chinese wine market through profiles of five over-achievers in wine-minded Ningxia.

Of China's six key grape-growing regions, autonomous Ningxia has the distinction of being heavily backed specifically as a wine producer by the local government.

Situated 683 miles (1,100 km) southwest of Beijing, it already has around 100 wineries (French heavyweights LVMH and Pernod Ricard have both invested here), and a vineyard area of roughly 40,000 ha (98,850 acres) which runs along the eastern foothills of Helan Mountain. The arid, sandy soils are planted with ungrafted vines (predominantly red wine grapes, often Cabernet Sauvignon, Cabernet Franc, Marselan, Merlot and Syrah, with Chardonnay and Riesling the most common white wine varieties).

Ningxia has enjoyed increasing media coverage since the 2012 launch of the Ningxia International Winemakers Challenge (a project that funds foreign winemakers wishing to hone their skills in the region), and the local government is pushing hard to promote the developing industry, as evidenced by their goal of 66,000 ha of vineyard by 2020, and the creation of the Asian Wine & Spirits-Silk Route conference, wine show and competition.

Held last month in the capital city of Yinchuan, the second edition of the Asian Wine & Spirits-Silk Route competition saw nearly 800 wines from across Asia (China, Georgia, Azerbaijan, Turkey, Moldova, Armenia and Israel). Chinese wines won 108 medals (out of a total of 264 medals) and of these 108 Chinese medals, 65 were won by Ningxia wines. The other Chinese medal-winning regions were Xinjiang, Hebei, Shandong, Gansu, Henan, Jilin, Liaoning, Neimengu, Tianjin and Beijing. The wines were judged by 49 tasters (including MWs Sarah Jane Evans, Annette Scarfe and Pedro Ballesteros Torres) from 18 countries. Ningxia wines did rather well: of the 18 Grand Gold medal winners, they scooped 11, and the region carried off 65 of the 132 medals awarded overall.

Many younger, wine-savvy Chinese people attended this year's Silk Route event as sommeliers, winemakers, wine educators and marketers, presenting themselves in fluent English using Western first names—although in my selection of five of the most impressive below, I have presented their Chinese names, surnames first as is Chinese custom. Their enthusiasm shone through and bodes well for the future of Ningxia wine.

Ren Yanling

Chief winemaker and production manager at Pernod Ricard's Helan Mountain,

Ren Yanling joined the estate as a cellar hand 18 years ago, saw it go from being government-owned, to a joint venture, to being fully owned by Pernod Ricard in 2012. Today, she runs a team of 20.

The daughter of a vineyard manager, 41-year-old Linda (the name she uses with English speakers) was raised locally but she has visited France, Italy, Australia and New Zealand (she worked the 2016 vintage at Church Road in Hawke's Bay). A trip to Bordeaux for the Vinitech trade show is scheduled for later this year.

When I met her, she was busy supervising harvest. Helan Mountain's 132 ha (325 acres) of vines are hand-picked (weed management is also done this way), which is typical for Ningxia where there is a plentiful supply of vineyard workers. This is just as well: the region's freezing cold winter temperatures (-5°F/-20°C is common) mean that the vine trunks have to be covered with soil from November to March.

However, Linda sees change ahead, due to an ageing labour supply combined with the area's urbanisation, and recent changes to China's one-child policy. Consequently, all new planting is trellised to make it machine-friendly: "We're preparing for the future."

The stated ambition for Helan Mountain is to increase production tenfold by the mid 2020s, so things are moving fast. More vines are being planted each year (Syrah, Marselan and Malbec are currently growing in the on-site nursery; in due course they will join the Chardonnay, Merlot and Cabernet Sauvignon already in production).

New offices were built in 2017, and the bottling line has been revamped. The original 32,000-litre stainless-steel tanks are complemented by smaller sizes ("we add new ones each year, so we can keep improving." explains Linda).

Since 2016, she has focused on wine quality and development, experimenting with the use of oxygen and whole-bunch pressing. She has also put new teams and management tools in place (charts detailing tasks and targets grace the winery walls, and a huge board displays plans for more building work).

On the topic of terroir, Linda remarks: "Ningxia wine has not yet got a uniform style. It's not the Old World style of France, nor the New World style of Australia. Some wineries here use traditional methods, others new ones. There are some common points, like deep colour and good structure. But there's still work to be done. Here at Helan Mountain, we're continuing to make improvements in aromatics."

The fact that Helan Mountain won a Great Gold, three Gold and one Silver medal in this year's Silk Route competition suggests that she's doing something right.

Chen Zhuyun

Zhuyun (or Erika, as she confidently presents herself in English) attended the Silk Route event to represent Ningxia producer Kanaan. A 31-year-old with a ready smile, she learned German in China before spending four years in the German Pfalz region studying viticulture and oenology, including a two-year work placement with Weingut Bergdolt and some judging experience at the Mundus Vini competition.

During her time in Germany, Erika organised presentations of Chinese wines for local enthusiasts and her fellow students: "they were totally surprised by both the quality and quantity of China's production—and they were also rather shocked by the price of most of the wines." she laughs (Kanaan's cellar door prices range from about €20 to €85).

In 2016 she returned to China where she now works as a wine promoter, educator and events organiser, running her own company out of a small city near Shanghai. Like many younger Chinese people, Erika is connected and comfortable with social media, citing its potential for the wine business. "It's especially important for marketing, for wineries who want to build their image and interact with consumers."

Of the multitude of apps available, she flags up WeChat (the popular messaging app with multi social functions including Moments, with a timeline feel similar to Facebook), Sina-Blog (a micro-blogging site used by many key opinion leaders, or KOLs as they are known in China) and Douyin (a newer app launched in 2016).

After tasting her way around the stands at the Silk Route show, Erika remarks on a tendency towards high alcohol levels and fruit-driven sweetness (she attributes this to the region's continental climate), and the prevailing use of oak.

However, she nuances her judgement. "It's challenging for me to taste these wines, as my palate has been used to German wine, which is fresher and more elegant, but I'm pleasantly surprised by what I've seen. China is a young wine-producing country, and I'm proud of what is being made here."

Li Hang

Steven (his English name) is 31 years old and the co-founder of the government-certified China Sommelier Academy. He studied hotel management in China, worked in Dubai and Abu Dhabi for five years (getting WSET level 3 and certification from the Court of Master Sommeliers along the way) and took a six-month sommellerie course in the US before returning to China.

The China Sommelier Academy trains 1,500 sommeliers, and brings around 450 students to Ningxia on field trips each year. It also creates Ningxia-specific teaching materials. Steven runs food-and-wine-pairing dinners on a regular basis, where he presents both imported and Chinese wines. "Ningxia wines are a must, to show that China can make very high-quality wine." he comments.

He believes that social media is key to wine promotion in China. "I uploaded a Sommelier Academy wine video this year and in the space of three days, it had 18 million views." The Academy's student recruitment is apparently mainly done through this channel.

Traditional Chinese medicine discourages drinking cold beverages (warm water is served with meals, and bottled beer is often drunk at room temperature), so when training staff to work in hospitality and retail, Steven spends time demonstrating the influence of service temperature so that staff are well-equipped to advise customers.

Speaking about food and wine pairing at the Silk Route conference, he mentioned that historically red has been the preferred choice for Chinese wine drinkers, but "because of the complexity of flavour in Chinese food and the way we tend to share dishes, white wine is often a more harmonious choice than red". He has seen a shift in drinking habits in the last three years: "Wine education is growing, consumers are better informed. White and sparkling wine consumption is growing in the first-tier cities, and based on feedback from trade shows, some importers and retailers are starting to place more importance on these products. In Beijing and Shanghai, for example, there are now companies that specialise in white and sparkling wine. I firmly believe that as the market continues to develop, these categories will grow in China."

Similarly, he has observed huge changes in Ningxia's wine production ("the quality has been continuously improving, there are now many estates which have won international awards") and he feels there is potential for this young region to become world-famous, although "it still has a long way to go, in terms of research into terroir and experimenting with new grape varieties, for example." He predicts that climate change may pose issues, as few wineries have sufficient experience to deal with this, and suggests that marketing is also an area in which progress can be made ("wineries need to start focusing on their branding").

Hurdles to be overcome include cost ("the initial investment to set up an estate is high and the degree of mechanisation is low, and these fine wineries have small output of around 50,000 to 100,000 bottles per year, so overall costs are relatively high. Compared with imported wine, our prices are not competitive") and consumer perception ("some Chinese people still believe foreign wines are better than domestic, regardless of price and quality").

Liu Aiguo

LVMH's Chandon winery is a 40-minute drive from downtown Yinchuan.

Liu Aiguo has been with Chandon since their Ningxia project was launched in 2012, joining the LVMH portfolio of sparkling wine production sites in Australia, Brazil, the US, India and Argentina.

Previously with Hansen Winery in Inner Mongolia, where he was winemaker and technology manager, he joined Chandon when he was 31 as Operations Winemaker, farming the company's 68 ha (168 acres) of Pinot Noir and Chardonnay to make 0.7 million bottles a year—the first time premium quality, traditional-method sparkling wine has been made in volume in China, according to him.

Alan has a master's degree in viticulture and oenology from China's Northwest A&F University (his wife was a fellow student: she went on to become an oenology lecturer and three of her former students have joined Chandon). He has visited France, Italy, Argentina, Brazil and California. Napa impressed him with its "excellent terroir, climate and top-quality wines", and he talks with enthusiasm about California's mature wine tourism industry.

He presented Chandon's original Brut and the newer Chandon Me (a play on the Mandarin word for "honey" or "sweet") in good English, explaining that the latter has been created as an easy-drinking bubbly with more fruity sweetness and gentle acidity to appeal specifically to the Chinese market. Although sparkling wine consumption in China is just one per cent of still wine consumption, Alan is convinced that bubbly has a future in his homeland: "The younger Chinese generation like new things, and they're more open-minded in their drinking habits. Sparkling wine represents the lifestyle to which these new consumers aspire."

Time will tell if he is right, but his belief is borne out by Wine Intelligence research, so the chances are on Alan's side.

Yang Weiming

Slightly built with elegant, long-fingered hands, 32-year-old Yang Weiming is softly spoken, and proficient in English and French. He is the winemaker at

Yuanshi, a stunning winery built from local stone designed to welcome tourists and wine professionals alike (it boasts attractive architecture and gardens, a long, dramatically-lit tasting table, smart overnight accommodation, a vast restaurant with a distinctive, rustic-chic décor and a wine shop).

After a degree in agricultural engineering and a one-year sommellerie course at France's Université du Vin in Suze-la-Rousse, winemaker Yang (Jonathan to his anglophone friends) joined Yuanshi in 2009 when building work had just been completed (the estate's first vintage was 2010).

As he prepared to make the ninth vintage from the estate's 809-ha (2,000-acre) vineyard, he commented: "In the beginning we wanted to make a bordelais kind of wine, because there was a lot of talk about Bordeaux. But as the years have gone by, we've evolved towards a style that reflects the Helan Mountain area. And we want to make wines that suit the Chinese palate: that's our aim for the future."

He realises this may be difficult: "In my work, the biggest challenge is the link between understanding Ningxia's terroir, and creating wines for the Chinese. I've spent many years using different techniques and oenological products to really express the potential of our grapes, and there's still a lot of work to do."

His time in France has given him a fondness for the wines of the Rhône Valley (he cites Châteauneuf-du-Pape and Vacqueyras as inspiration). He relishes the struggle to make fine wine—"the harder you try, the more challenges you face, so I never get bored"—and despite the Chinese market's marked preference for barrel-aged reds, Yang makes a fine, unoaked Chardonnay. He muses: "Our region's wines need more time to develop and find their place here in China first, and then overseas. We have to take all available opportunities to make our wines known outside of Ningxia, and welcome visitors who want to come and discover them here."

China Daily USA
Ningxia Wines Wow UN Diplomats

usa.chinadaily.com.cn/a/201811/09/WS5be46b6ca310eff3032877ca.html

By Hong Xiao
9 November 2018

They raised their glasses to the wines of Ningxia at United Nations headquarters in New York this week.

As part of the Chinese Food Festival, which runs from Monday to Friday this week, more than 20 UN ambassadors and diplomats attended an exclusive luncheon on Wednesday.

Cuisines for the luncheon were prepared by chefs from the Ningxia Hui autonomous region, and international award-winning wines produced in the eastern foothills of the Helan Mountains in the Northwest China region were featured.

Ravi Batra, adviser to the Permanent Mission of Ukraine to the UN, shared a story about Chinese liquor during the luncheon. He recalled the time that India's ambassador was hosting a Chinese diplomat, and Maotai, the most renowned Chinese baijiu, was served.

"It scared me that nothing had never gone down off my palate that is stronger than Maotai, but today, it (Ningxia wine) changes my view; this is so wonderful!" he sighed after tasting the wine.

Cao Kailong, director of the Ningxia Grape Industry Development Bureau, told the guests that the wine was made from grapes planted in the eastern foothills of the Helan Mountains in Ningxia located between 37 and 39 degrees north latitude.

"It is believed to be the 'golden zone' for wine-grape cultivation in the world," he added.

"The soil, sunshine, temperature, precipitation, altitude, hydrothermal coefficient and other conditions constitute a perfect combination to produce grapes with well-developed aroma, good chromogenesis and a balanced acid-sugar ratio," Cao explained.

The wine region of the Helan Mountains in Ningxia has 86 wine chateaus and plants nearly 40,000 hectares of wine grapes, which can produce nearly 100,000 tons of wine per year. It is the only chateau wine-producing area in China.

"It has become China's most promising wine-producing area on a par with the world's best wine-producing areas," Cao said confidently.

Alexandru Cujba, former permanent representative of Moldova to the UN, recalled his visit to the Ningxia region last September.

"I was deeply impressed by what I have seen over there. This region is remarkable. The results achieved by Ningxia in the area of agriculture, particularly the grape and wine industry, are extraordinary," he said.

"China is an example that there is kind of a global awakening of thoughts on wine that more and more people are going into wine(making)," said Kaha

Imnadze, permanent representative of Georgia to the UN.

Imnadze said that there's competition in most industries, "but in wine, the more people go into wine, it's better for every winemaking country".

The great advantage that Ningxia has is its variety and potential as a winemaking region that is becoming globally known.

"People tend to look for new wines, because they want to indulge themselves in different wines, not necessarily wine coming from a big name, but quality wine may come from a chateau."

Imnadze, who has been to China several times as a tourist or a government guest, said he has seen the success that China has achieved in many fields.

Imnadze said his country, Georgia, is a cradle for winemaking and has more than 500 grape varieties. "And not all of them are fully commercialized," he said.

He invited the Ningxia delegation to visit Georgia to promote exchange and cooperation in grape planting and winemaking between the two countries.

"You can be a leader in producing and giving the world some of the newest tastes that many do not know about," he said.

The 2018 Chinese Food Festival is co-organized by the Human Health Organization, the Administration of Development of Grape Industry of Ningxia and the UN Delegates Dining Room.

The festival features authentic Chinese food, blending traditional Chinese flavors, such as imperial court dishes, and regional cuisines from Ningxia.

The Drinks Business
China Launches Ningxia Wine Train from Beijing

thedrinksbusiness.com/2018/08/china-launches-ningxiawww.thedrinksbusiness.com-wine-train-from-beijing/

By Natalie Wang
21 August 2018

China has rolled out a wine train connecting the country's capital Beijing to the wine producing region of Ningxia in northwestern China, offering passengers the chance to taste wines, try out wine-infused facial masks and gain first-hand harvest experience at wineries. (This is an excerpt. Use the url at the start for the full article.)

Just-Drinks
Why the World's Wine Producers Won't Have It All Their Own Way in China

just-drinks.com/comment/why-the-worlds-wine-producers-wont-have-it-all-their-own-way-in-china-comment_id126945.aspx

By Chris Losh
11 October 2018

Most established wine producers seem to have assumed that the Chinese would sit passively by while they flooded the country's shelves and reaped the benefits, particularly above the RMB100 (US$14.50) per bottle level. The rise of the Ningxia region, however, proves that France, Australia et al have a fight on their hands.

2019

Wine-Searcher
China's Wine Egg Hatches

wine-searcher.com/m/2019/04/chinas-wine-egg-hatches

By Jim Boyce
28 April 2019

A new winery development in Ningxia is impressive, but Jim Boyce points out that it's a gamble.

An egg-shaped complex called Xige in Chinese, and known informally as Pigeon Hill in English, has risen from the dusty fields of Ningxia in north-central China. Residents of those windswept plains have witnessed inspiring blueprints transform into a sprawling winery nestled within a curved stone wall during the past two years. And now the first wines are set for release.

But Xige (She Guh) is more than just another fancy winery. The Ningxia region's government hopes it might be a kind of Penfolds, a respected high-volume national brand that makes everything from palatable entry-level wine to the finest reserve cuvées. And, by doing so, that it can help the rest of the wine industry, because birds of a feather flock together, right?

The project exposes a challenge faced by Ningxia and the China wine sector at large. In less than a decade, Ningxia went from virtual unknown to aspiring virtuoso, winning more than a thousand medals and kudos from the world's leading gatekeepers. But that critical success was not fully reflected in sales and meant pressure for both the government and industry.

To be fair, this isn't all Ningxia's fault. Long before the region joined the wine scene, bigger producers set the stage for a credibility crisis by leveraging consumer ignorance and stressing marketing over quality. When quality finally did start to rise, many consumers, who then knew a thing or two about wine, were turning up their noses at local brands in favor of imported ones. The chickens had come home to roost.

Can a 25,000-square-meter winery help scare them off?

To be sure, Xige has plenty of bells and whistles. It controls the biggest and oldest swath of vineyards in Ningxia, with some 1,000 hectares, mostly Cabernet Sauvignon, dating to the 1990s. There are also 300 hectares of newer plantings, including of Malbec and Marselan.

Those grapes go to a modern facility with 10 million liters of capacity and loads of top-notch equipment, from a German press to a VinWizard tank control system from New Zealand, to the thousand barrels used for the 2017 vintage alone.

On top of this is a team of local and international experts.

The key here is owner Zhang Yanzhi, who poses a double threat as both Bordeaux-trained winemaker and head of importer and distributer Easy Cellar, which already handles major brands. This distribution network might well be the X-factor for Xige, given that finding a route to market has proven tough for others. Zhang also has financial support: according to local media, Xige initially had funding of $40m, raised with support from Ningxia's Wine Trading Expo Center.

Zhang's team includes chief winemaker Liao Zusong, who worked at Grace Vineyard, one of China's best operations, with stints at Bass Philip and Mollydooker in Australia. Also on board as consultants are Bruno Vuitennez and Valerie Lavigne, while Christelle Chene has joined as export sales director.

It sounds like all the ducks, er, pigeons are in a row as the first wines are set to roll. The initial three-tier wave includes entry-level N28. It's also called Helan Hong, named after Ningxia's mountain range, and is one of the wines endorsed by government to represent the region. (If you attend official functions, you might down a few glasses.) Given the Penfolds analogy, N28 is not exactly Rawson's Retreat as it will retail at $24, but I'm told less expensive options will be coming.

One level up is N50 Old Vines, made with fruit from those 20 year-plus vineyards. And next is the Jade Dove series, including single-vineyard Chardonnay, Cabernet Sauvignon and Cabernet Gernischt.

The top tier, Xige Reserve, is still maturing in barrel.

Most of the wine should be siphoned into Easy Cellar's network, with some going to corporate and hoped-for export clients.

Oh, and Xige's hotel should be ready in a few weeks, making it easy for guests to guzzle wine and then nap it off.

Xige is nothing if not ambitious, but that also brings risk. What if this blend of high-end equipment, vineyards and consultants, with strong government, financial and distributor support, doesn't meet expectations? What then? And what if the wines are a success but Xige's superior scale undercuts smaller wineries rather than boosts them?

All this makes Xige somewhat a gamble. Then again, much of Ningxia's wine development could be seen in that light.

In any case, the wheels are in motion. Xige samples were presented at the huge annual wine show Tang Jiu Hui in Chengdu in March. Xige was also part of a multi-city tour, led by Jeannie Cho Lee and Ma Huiqin, to promote Ningxia wine. And the winery has hosted a steady flow of visitors, including a slew of officials, as it prepares to take center stage.

Interestingly, Xige is not in the shadows of the famed Helan Mountains, home of the region's best-known wineries, such as Silver Heights, Legacy Peak, Helan Qing Xue, Kanaan and Chandon. Instead it sits further south, past the tail end of that range, where the wind is stronger and winter harsher.

The vines here have a unique history, a maverick one. If they could talk, they might tell of once being part of a joint venture with Pernod Ricard. Or, more recently, of providing grapes for Ningxia Winemakers Challenge, in which 48 contestants from 17 nations competed in a $100,000 winemaking contest.

Or of a story rooted much deeper in the past. In 1997, China's President Xi Jinping visited Ningxia as a Fujian province official on a poverty reduction mission. One result was to shift people from the harsher areas of Ningxia to those not far from Xi Ge. And one sector that has since arisen in those parts is wine. In fact, Xi Ge's first vines were planted the same year as Xi Jinping visited. Those who believe in fate might see this as another X-factor that will decide the project's success.

Imbibe Magazine UK
Enter the Dragon: How Ningxia Is Putting Chinese Wine on the Map

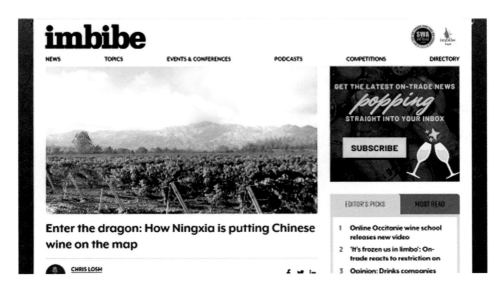

imbibe.com/news/enter-the-dragon-how-ningxia-is-putting-chinese-wine-on-the-map/

By Chris Losh
14 June 2019

Ningxia has established itself as China's premier fine wine region in little more than 10 years. Chris Losh heads up to Helan to see whether this fast-growing region really could be the Chinese Napa.

Lenz Moser puts down his refractometer and gazes at the recently harvested Cabernet vineyards in front of him.

"You know, China thinks long term, but it makes things happen in the short term,' the Austrian winemaker says. 'The pace of change in this country is truly amazing."

He's not wrong. We're on the edge of the city of Yinchuan, a two-hour flight west of the Chinese capital, Beijing. Thirty years ago, this was a small town of 30,000 people. Now it's a city of two million and the government has plans to double its population to four in the next 10 years. In every direction eight-lane highways stretch into nothingness, ready for the soon-to-be-built high-rises.

Yinchuan is the biggest city in Ningxia (pronounced Ning-shah) province. And if there's one word that you need to learn about Chinese wine, it's this one. Ningxia is not about volume—it's responsible for only about 10% of the country's production. But its focus on quality wines means it punches above its weight and there's an unmistakable buzz to the region—a feeling that this could be the place to put Chinese wine on the map.

Putting down roots

Moser might be Austrian by birth, but he's made wine all over the world, including California, and he likens the energy here to Napa Valley in the 1970s. This, of course, is when the legendary Californian vintner Robert Mondavi was travelling the world telling anyone who'd listen that his wines deserved to be judged against the finest in Europe—worth remembering should you be tempted to dismiss Chinese wine as being too niche to bother about.

Having worked with China's biggest wine company Changyu for over a decade, Moser has partnered with it to create a new estate in Ningxia. Château Changyu Moser XV is a vanilla-coloured slab of rococo grandeur straight out of the Bordeaux playbook, and its ambitions are obvious.

Moser has spent much of the last three years schlepping his bottles round Europe and has managed to secure listings and importers throughout the continent.

"First we have to surprise people, which I think we are doing," he says. "Then we want to keep on getting better."

For Moser, 2015 was a watershed vintage. It was a good one for starters, after a trickier 2014, but his vines were also 10 years old by then, meaning they were "starting to get interesting".

Breaking ground

The sheer youth of the vines is a reminder of just how embryonic the wine industry is here—and, like the city itself, of how fast it's grown. The first winery of any description appeared here in the mid-80s, with the smiley, chatty Jing Zhang setting up the first boutique one 20 years later.

Visit her Helan Qingxue winery—the name means "on a sunny day you can see the shining snow"—and there are photos of bulldozers laying the groundwork in what was essentially a vast desert. The pictures are from 2005. Just six years later, Jing's Jia Bei Lan Cabernet won International Wine of the Year in the Decanter World Wine Awards.

Such international approval opened the floodgates. Already Ningxia has 38,000 ha under vine—about the same as New Zealand—and over 100 wineries. The next 10 years should see the vineyard area increase to 60,000 ha.

To describe the pace of change here as "frenetic" barely does it justice. There's no shortage of space—Ningxia is one of the least populated regions in China—and vines and wineries are springing up everywhere.

This is how your correspondent found himself in a car being reversed at speed down a white dust track.

We can see He Jin Zun, the new winery we want to visit, we just can't get to it. A year ago, the place didn't exist so no one knows the directions, and signposts haven't been put up yet because the roads haven't been finished. Nor, it turns out, has the winery.

"They are literally building the winery as we go," says Dave Tyney, the affable Kiwi winemaker at He Jin Zun. "Two weeks ago, this floor was dirt and there were no tanks. Now they put in a tank and we fill it the next day. It's stressful, but you get used to it."

Ningxia and Marlborough are officially "friendship regions" and there's no shortage of Kiwis working among the semi-built chaos. It's hard to imagine anywhere on earth less like Marlborough, which, for the young team mainlining coffee and frantically scribbling on pads of paper, might be part of the attraction. Ningxia has its rough edges and quirks, but it's undeniably exciting.

Method or madness

Look on a map and you'll see that Ningxia is on the same latitude as Napa Valley. But the two could hardly be more different. While Napa is all about valleys and sea fog, Ningxia is essentially a high desert.

A huge, open dusty plain, Ningxia's vineyards are high at 1,100m above sea level and 1,000km from the nearest sea. Hot in summer and freezing in winter, the climate is about as continental as it gets. In the shadow of the towering Helan Mountains to the west, rainfall is miserly—just 200mm a year—though most of it falls, unhelpfully, in July and August.

Other than that, it's an excellent climate for grape growing. Even in the height of summer, there are no heat spikes. And since the temperature rarely gets above 35°C the vines don't shut down.

It reminded this journalist of Mendoza, where similarly the altitude creates a big diurnal shift that preserves natural acidity in the grapes. Temperatures typically drop below 20°C at night in midsummer, while closer to harvest they are already in single figures.

This rapid fall in the mercury is one of Ningxia's peculiarities. Winters are brutally cold at -20°C, not including wind chill, and come in quickly. Since vines die below -18°C, the Chinese growers have a unique solution to prevent this happening—they bury them.

After vintage, the plants are quickly pruned, then the trunks laid down and covered with earth to create what looks like rows of Saxon burial mounds. The growers then flood the earth so that it freezes. Encased in ice, the plants are thankfully protected from the potentially fatal, super-low temperatures of mid-winter.

The need to bury the vines at the end of the growing season impacts all the way through viticulture in Ningxia. The rows, for instance, need to be wide enough to get a tractor down them to plough up the soil, which means no high-density plantings.

Plants have traditionally been trained in the 'single dragon trellis', which makes burial easier, but is not always good for maximising grape ripening. The burying takes its toll on the vines, too.

Since vines need to be picked by mid-October and pruned by the start of November (to ensure safe burial by 20 November), this is not somewhere that growers are tempted to leave grapes on the vine for an extra couple of weeks in search of a few more degrees Baumé.

"The biggest challenge we have is the short growing season," says Yang Weiming, winemaker at the strikingly beautiful Château Yuanshi winery. "Frost can come at the end of September. If we want freshness we really have to pick early, but phenolic ripeness comes late."

Weiming is one of a plethora of bright young Chinese winemakers driving the Ningxia revolution. Both highly qualified and well travelled, all of them seem to have worked in France following their degree, and many have also put in time in other parts of Europe and (albeit less commonly) the New World.

Rather than a love of fruit ripeness, they've brought back with them a European appreciation for structure and freshness in their wines. Silver Heights' Emma Gao Yuan, who spent several years working at Calon Ségur in Bordeaux, also returned with a French husband.

"I want to bring French taste here, not just ripe grapes," she says. "I prefer the less-perfect years like 2014, which are cooler and lower in alcohol."

Weiming's point about phenolic ripeness, however, is well made. If there's a common thread through the wines, it's of lifted fruit followed by squeaky tannin; of wines that start with elegance, but get a touch green on the finish.

This may partly be down to the price that the winemakers are prepared to pay to chase freshness, but it's just as likely to be a function of the viticulture. Yields per hectare might look fine, but take into account the rows being so far apart to facilitate that all-important winter burial and the production per plant can actually be relatively high.

Likewise, the undoubted expertise in the wineries is not yet matched in the vineyard. Canopy management techniques, which could slow down ripening to allow phenolic ripeness to catch up with sugar ripeness seem to be practised rarely, if at all.

At Château Changyu Moser XV, hiring a local vineyard manager is high on Moser's list of priorities. "I want to work with Chinese people—we are in China!" he says. "But [in the vineyards] it's more difficult. The universities here put the focus more on winemaking than viticulture."

A new dawn

This, perhaps, is where the Ningxia revolution comes up against its greatest challenge. After all, you can learn how to make good wine in a few years, but understanding what to plant and how to get the most out of your vines can take centuries. And here the industry appears split between those such as Qing Song Xao, owner of Lan Cui winery, who asks "why should we copy Napa or Bordeaux?", and those such as Lenz Moser who wonder "why reinvent the wheel?".

Currently, for sure, Bordeaux's influence looms large, with most wineries intent on a Cabernet-based château model that plays well with the local market. Vinously, as well as politically, reds dominate. White wines are barely 10% of the production.

Cab is a variety that clearly has potential here, though in the opinion of this journalist, seems to perform better when it has a little Merlot in the mix to plump out the mid-palate and ease the tannins.

But you can't help but feel that the uniqueness of this region means there ought to be another variety that will take off here and really put Ningxia on the map, an equivalent of Malbec in Mendoza or Sauvignon Blanc in Marlborough.

A few years ago, there was a lot of talk about Cabernet Gernischt (Carmenère) which, given how suicidally late it ripens, seems positively masochistic in a region prone to autumn frost. Many wineries have experimental vineyards, and everything from Tempranillo, Zinfandel, Muscat, Mourvedre, Syrah and Riesling are being trialled at the moment. But the variety that came up most often in discussions with winemakers was the French red-wine grape Marselan. A French Cabernet/Grenache cross, it has potential and some wineries, such as Domaine Pu Shang, are able to attract seriously high prices for it.

That said, neither the Bordeaux blends, nor the Marselans, have a definable "Ningxia style" yet. That may well come with more vine age and better viticultural knowledge. It's still very early days.

So, will the UK restaurants be awash with Ningxia wines in the next couple of years? Probably not. But there are a small number of good producers who are making an effort to expand internationally, and their wines have real merit. Jia Bei Lan is already here (Panda Wines), as is Château Changyu Moser XV (Bibendum), and I'd expect Silver Heights to be snapped up shortly.

It's early days, but as a famous Chinese proverb points out, "A journey of a thousand miles begins with a single step."

Indian Wine Academy
3RD Edition of BRWSC Highlights Ningxia as Top Wine Region of China

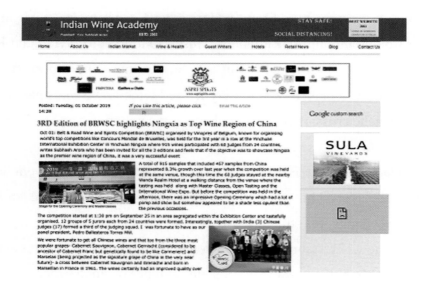

indianwineacademy.com/item_6_824.aspx

By Subhash Arora
1 October 2019

Belt & Road Wine and Spirits Competition (BRWSC) organised by Vinopres of Belgium, known for organising top world competitions like Concours Mondial de Bruxelles, was held for the third year in a row at the Yinchuan International Exhibition Center in Yinchuan, Ningxia, where 915 wines participated with 60 judges from 24 countries, writes Subhash Arora, who has been invited for all three editions and feels that if the objective was to showcase Ningxia as the premier wine region of China, it was a very successful event.

The total of 915 samples included 457 samples from China and represented 8.3% growth over last year when the competition was held at the same venue, though this time the 60 judges stayed at the nearby Wanda Realm Hotel, walking distance from the venue where the tasting was held along with the master classes, open tasting and the International Wine Expo. But before the competition was held in the afternoon, there was an impressive opening ceremony which had a lot of pomp and show but somehow appeared to be a shade less opulent than the previous occasions.

The competition started at 1:30 pm on September 25 in an area segregated within the exhibition center and tastefully organised. Twelve groups of five jurors each from 24 countries were formed. Interestingly, together with India (3), the Chinese judges (17) formed a third of the judging squad. I was fortunate to have as our panel president, Pedro Ballesteros Torres MW.

We were fortunate to get all Chinese wines and that, too, from the three most popular grapes-Cabernet Sauvignon, Cabernet Gernischt (considered to be an ancestor of Cabernet Franc but genetically found to be like Carmenere) and Marselan (being projected as the signature grape of China in the very near future), a cross between Cabernet Sauvignon and Grenache and born in Marseillan in France in 1961. The wines certainly had an improved quality over the last year's tastings at the same venue.

For the statistically inclined, there were 17 countries that participated, with the most samples (457) from China followed by 42 from Moldova and 31 from Greece. As might be expected, 599 samples were red and 251 were white—this seems to be a higher ratio than the white wines consumed in China. Sixty-five of the 915 samples were rose, again a seemingly higher percentage than what is actually drunk—perhaps there is a new trend. They were vying for the Gold Prize (95-plus points), Grand Prize and Silver Prize.

The competition was a continuation of China's Belt & Road Initiative under which the Asia Silk Route Forum and Tasting was founded in 2016 in the Fangshan wine district of Beijing. After the successful competition in 2018, it returned for the third time this year on September 25-28 in Yinchuan, as the region has the largest concentrated production area of wine grapes in China and shows great potential. More importantly, it has been marketed well by the local government authorities with the financial support of the central government. The theme chosen for the current event was Identity, Dialogue and Integration.

Masterclasses

There were a few masterclasses held on September 26... These included a presentation in Chinese on wines and finance by Kangren Chen, the Vice Mayor of Yinchuan. This was followed by a short presentation by Zara Muradyan, Executive Director of the Vine and Wine Foundation of Armenia on Armenian wines, followed by Giorgi Samanishvili, oenologist and consultant, with a guided tasting of five wines from Georgia. Bernard Burtschy made a good case for organic viticulture being less difficult and important in China.

The highlight was a presentation by Professor Demei Li, who is by now a familiar face at any Chinese symposium. One of the top ten wine consultants of China, who literally brought the Ningxia region to the attention of the world when one of the wineries he consults for in the eastern foothills of the Helan Mountains—Helan Qingxue Vineyard, whose Grand Reserve Jia Bei Lan 2009 won the trophy for the best international wine over £10 in 2011, spoke about wines in China in general and Ningxia in particular, with a guided tasting of five wines.

Winery Visits and Tastings

Besides the open tasting for three days at the International Wine Expo at the exhibition center where over 200 producers from 20 countries showcased their wines giving the visitors an opportunity to taste several wines from China and in particula Ningxia, the judges were invited to several wineries during the hectic schedule...

The event has been conceptualised and partnered by Concours Mondial with Beijing International Wine and Spirit Exchange, which we had visited in 2016...

One thing for sure is that Ningxia will roar like a lion in the near future. The government of China and Yinchuan is betting on the right horse. The prices are still high but good wines should be available at better prices soon. Otherwise, the charm of Ningxia wines may be lost—at least in the international markets.

Bettane + Desseauve
Helan Hong, Nouvelle Marque de Vin Chinoise

mybettanedesseauve.fr/2019/10/08/helan-hong-la-nouvelle-marque-de-vin-chinoise/

By Par Mathilde Hulot
8 October 2019

Pendant que les festivités du 70e anniversaire de la République Populaire de Chine battent leur plein, le gouvernement autonome de Ningxia, appuyé par le grand patron Xi Jinping, lance une marque de vin qui portera le nom de Helan Hong.

Une nouvelle appellation bientôt plus grande que le vignoble bordelais.

"Helan" pour la chaîne de montagnes qui domine le vignoble de Ningxia, situé à un peu plus de mille kilomètres à l'ouest de Pékin, et "Hong" pour rouge, la couleur du bonheur et de la réussite pour nos amis chinois. Il manquait, en effet, à côté des Great Wall, Dragon Seal et autre Shangyu, un nom fort capable de hurler au plus grand nombre l'existence de cette appellation toute neuve qu'est le Ningxia. Cette région viticole, à peine sortie du nid puisque ses "vieilles vignes" n'ont que vingt ans, a poussé comme du bambou. Le gouvernement fédéral et local en ont fait le vignoble phare du renouveau gastronomique chinois. Situé aux portes du désert de Gobi, il fait aujourd'hui 40,000 hectares et pourrait supplanter en surface notre bon vieux vignoble de Bordeaux d'ici quelques années. Un jeu d'enfant pour ce peuple que rien n'arrête, ni la chaleur, ni le froid (il faut enterrer les vignes en hiver durant lequel le thermomètre peut descendre à-30°C) ni la sécheresse, ni la mévente. Seul obstacle tout de même: le fleuve Jaune qui, très sollicité, peine à arroser ces surfaces gigantesques et artificielles.

Helan Hong, qui sonne comme un gong, a vu le jour dans la tête des dirigeants fin 2018. Une marque ombrelle, qui verra sans doute différents produits sous ce nom. Pour l'instant, il s'agit d'un cabernet-sauvignon. Quatre entreprises sont habilitées à le produire, à partir d'un cahier des charges défini : He Jin Zun, Ho-Lan Soul, Ningxia Pigeon Hills et l'Imperial Horse Winery. Une cinquième cave est en construction sous l'égide de l'État. Actuellement, 100,000 bouteilles seulement sont accessibles, dont 3,000 d'une cuvée Réserve destinée à nos amis Belges qui fêtent en ce moment-même l'arrivée de ce vin plaisant, d'une belle fraîcheur, très représentatif de la jeune appellation, au parc d'attraction Pairi Daiza et au centre de Bruxelles.

Ce Helan Hong revu à l'Européenne (pour que nous puissions en déchiffrer le nom) est le cheval de Troie idéal pour pénétrer nos marchés occidentaux. Les Chinois comptent, à terme, produire 20 millions de cols de ces "montagnes rouges", dont "seulement" 1% à l'export. La nouvelle marque compte bien rendre plus populaire les vins de Ningxia qui, il est vrai, sont encore actuellement bien chers pour le commun des mortels, entre 25 et 150 euros la bouteille.

Vitisphere
Chandon Lance Son Premier Vin Rouge Pétillant de Chine

vitisphere.com/actualite-90590-Chandon-lance-son-premier-vin-rouge-petillant-de-Chine.htm

By Alexandre Abellan
13 November 2019

Lancé en septembre dernier par le pôle vin du groupe Louis Vuitton Moët Hennessy (LVMH), la cuvée Xi vise dans un premier temps le réseau des cafés, hôtels et restaurants (on-trade). L'ambition de ce nouveau produit étant de créer de nouveaux modes de consommation en Chine, permettant d'ouvrir une nouvelle voie d'accès aux vins effervescents.

2020

The Washington Post
These Are the Women Leading China's Wine Revolution

washingtonpost.com/photography/2020/01/24/these-are-women-leading-chinas-wine-revolution/

By Matilde Gattoni & Matteo Fagotto
24 January 2020

Ningxia is the hidden gem of north-central China. Looking at its majestic peaks and orderly vines stretching as far as the eye can see, it is difficult to imagine that a little more than 20 years ago, the region was just a forlorn stretch of sand inhabited by subsistence farmers. "When I was a kid, I used to dig holes in the desert. I would hide there and play together with my friends," remembers Ren Yanling, 42, while sitting in her laboratory during a rare break...

Sharp and confident, with a characteristic penetrating gaze, she is part of a group of strong-willed, talented female owners, winemakers and managers who are leading China's wine revolution. (This is an excerpt. Use the url at the start for the full article.)

The Drinks Business
Ningxia Region Continues to Grow

thedrinksbusiness.com/2020/04/ningxia-region-continues-to-grow/

By Alice Liang
9 April 2020

Ningxia is one of the renowned winemaking regions in China and, despite the Covid-19 pandemic, the development of the local industry has not stopped.

As with so many other regions in the world, the wine industry in Ningxia has been greatly impacted by the Covid-19 epidemic.

As such, the Ningxia Agriculture and Rural Department and the Finance Department has announced a fund of 30 million RMB to subsidise the industry as many wineries are expected to encounter difficulties in their cash flow.

Moreover, in the first quarter of 2020, six ventures in Qingtongxia, Hongsibao and Helan submitted plans for the building of five new wineries and the expansion of an existing property. The total investment is expected to be more than 250 million RMB and the total wine production volume will increase by 2,000 tons or more.

The eastern foothill area of Helan Mountain in Ningxia is recognised as one of the up-and-coming sub-regions due to its ideal terroir with plenty of sunshine, sandy soil, low precipitation and continental climate.

Recently, the Management Committee of the Shandong Lu Grape Industrial Park in Helan, Ningxia revealed that as of the end of 2019, the Ningxia grape planting area had reached 570,000 mu (equivalent to 38,000 hectares), accounting for a quarter of China's plantings, which is the country's largest contiguous wine grape production area. Some of the widely planted grape varieties in the region are Marselan, Malbec, Pinot Noir, Riesling and Chardonnay.

One of the significant features of their viticulture is that they need to bury the vines to protect them from harsh winter temperatures which can fall as low as -25℃. In the beginning of March, it was reported that the vineyards had started unearthing and fertilising the vines. The process is expected to be complete by mid-April.

Currently, there are 211 wineries with an annual output of 130 million bottles of wine, and the value of the region's winemaking industry has reached 26 billion RMB. Some of the boutique wineries, for example, Helan Qingxue and Silver Heights, in the Yinchuan sub-region have successfully made a name for themselves worldwide.

Big international players such as LVMH and Pernod Ricard have also invested in Ningxia and founded the Domaine Chandon and Helan Mountain wineries.

The region has been actively promoting its wine industry in recent years, generating 120,000 job opportunities. Wine tourism is another one of the key focuses as the region is receiving over 600,000 tourists annually.

People's Daily
Grape Wine Industry Turns Ningxia's Barren Land into Prosperity

en.people.cn/n3/2020/0811/c90000-9720281-2.html

By Yan Ke & Wang Hanchao
11 August 2020

The eastern foot of Helan Mountain, on the outskirts of Yinchuan, capital of northwest China's Ningxia Hui Autonomous Region, is a contiguous area of wine production, which is hailed as a "purple business card" of the Hui autonomous region.

From the first grape seedling planted some 20 years ago, to building a complete industrial chain and influence, the grape wine industry is thriving on this once barren land.

Since 1996, China made the collaboration on poverty alleviation between the eastern and western regions a major national strategy. Xi Jinping, who then served as the deputy secretary of the Communist Party of China (CPC) Fujian Provincial Committee, was in charge of Fujian's effort to assist Ningxia. Fujian-Ningxia collaboration officially set off since then.

Ningxia's Xihaigu region was once listed as the "most unfit place for human settlement" by the United Nations for its barren land and lack of water resource. Fortunately, a resettlement township was founded on the sands of a Gobi desert south of Yinchuan, capital of Ningxia, in 1997, absorbing over 40,000 residents from Xihaigu. To commemorate the close bond of collaboration between Fujian (also known as Min for short) and Ningxia (also known as Ning for short), the township was named Minning.

Poverty alleviation was carried out simultaneously as Minning started construction and resettlement work, and the grape wine industry was an important pillar for the town to shake off poverty. Entrepreneurs from Fujian believe that the Gobi desert where Minning is located is not an average one. Situated 38.5 degrees north of the Earth's equatorial plane, Minning sees low precipitation, long sunshine hours, and high diurnal temperature variation. Besides, the rich minerals in its high-permeable sandy soil offer great growth conditions for wine grapes.

With capital and technical input, the Gobi desert of Minning was soon covered by green leaves and filled with the aroma of wine. Thanks to the grape wine industry, the relocated residents from Xihaigu have had stable jobs and increasing income. As a result, they stayed in Minning.

Yang Cheng resettled in Minning after living in the mountains for decades, where he planted potatoes and wheat and worried all day for irrigation. Since he moved to the new town, his family all started working in a vineyard. Yang became an electrician after training, and his wife a sanitation worker. Plus, his son drives excavators for the vineyard. The family of three earns over 10,000 yuan ($1,435) per month.

The vineyard where Yang's family works is operated by Fujian businessman

Chen Qide. Starting on a 6,667-hectare wasteland, Chen vowed to make the best wine in Ningxia. Now, the wine he makes wins international awards almost every year.

As of the end of 2019, the eastern foot of Helan Mountain had a total of 38,000 hectares of vineyards, offering around 120,000 jobs for relocated residents.

Poverty alleviation of Minning started from wine making, but never ends with it. Fujian-Ningxia collaboration also includes constructing reliable sales channels and marketing methods.

Lai Youwei is a cadre from Dehua, Fujian designated to serve a temporary position in Minning. After arriving in Ningxia, he has invited many business people from his hometown to the vineyards in Minning, recommending products to them to expand the sales channel. He also joined a livestream marketing show this May to help sell the wine during which he finished orders of nearly 300,000 yuan.

Zhu Wenzhang, from Jinjiang, Fujian, is a wine dealer now working in the eastern foot of Helan Mountain. He introduced an innovative method of "shared winery" based on his connection with wine dealers who prefer high-quality grape wines. The method helped more than 50 enterprises "claim" a total of 200 hectares of vineyards as the origin of their production, so that the quality is better controlled. Besides, grape growers don't have to worry about the sales.

The grape wine industry witnessed the efforts and results of Fujian-Ningxia collaboration over the past 20 years, and helped many families realize their dream of getting rich. Today, Fujian-Ningxia collaboration is still upgrading, aiming to achieve more splendid targets.

Beijing Review
East-west Cooperation Program Facilitates Development of Poverty-stricken Region

bjreview.com.cn/Nation/202008/t20200810_ 800217158.html

By Lu Yan
13 August 2020

Recently, Zhang Junning, 31, bought a car. This was unimaginable six years ago, when his family was still living in poverty.

At that time, he had just moved from Xihaigu, an impoverished mountainous region in Ningxia Hui Autonomous Region in northwest China, to Minning Town some 300 km away.

Away from the previous life of traditional farming, he was at a loss in his new surroundings in the beginning. All he could do was odd jobs like working on construction sites and he could barely make ends meet.

But the pressure was soon relieved as he and his wife were hired by a vineyard. After receiving training and self-learning, they gradually gained grape-growing skills and a stable and decent income.

Seeing more and more villagers like him being hired by the vineyard, Zhang came up with the idea of establishing a labor agency with the money he had earned and helping vineyards hire local villagers.

His entrepreneurship enabled him to not only shake off poverty, but also build a new house and live a comfortable life.

"If it weren't for the national poverty alleviation program, I might still be a poor farmer who wouldn't be able to achieve this," Zhang told China Central Television.

Fruits of collaboration

Since the mid-1990s, paired cooperation between China's more developed eastern regions and less developed western regions has been a major part of the strategy to alleviate poverty and bridge the previously widening wealth gap between these regions.

At a conference on poverty alleviation cooperation held by the State Council in May 1996, 10 developed provinces in east China were paired with 10 underdeveloped regions in the west. Fujian, a coastal province in southeast China, was chosen to aid Ningxia.

In Ningxia, there are some areas like Xihaigu that were in poverty mainly due to extreme weather. "Droughts in summer, floods in autumn and freezing cold in winter" is how locals describe the region. In 1972, the United Nations classified Xihaigu as one of the world's most uninhabitable places.

Under the Fujian-Ningxia partnership, a resettlement program was proposed to move entire village communities from impoverished areas like Xihaigu to more fertile land near the Yellow River. Minning is one of the resettlement locations.

As part of a poverty alleviation scheme, the development of Minning started in 1997 based on the idea of Xi Jinping, who was then deputy secretary of the Fujian Provincial Committee of the Communist Party of China and in charge of Fujian's efforts to assist Ningxia.

The name Minning denotes the partnership between Fujian and Ningxia, as min refers to Fujian, while ning stands for Ningxia.

Now there are over 60,000 residents in Minning. The per-capita yearly disposable income of residents in Minning rose more than 20 times to a little under 14,000 yuan ($2,013) last year, according to Zhao Chao, the town's deputy Party chief.

Zhao said the town will ensure that all 22 remaining poor households shake off poverty, as 2020 is the finish line to achieve the target of eliminating absolute poverty and building a moderately prosperous society in all respects.

The villagers who did not want to move to Minning were not left behind. The local government has supported the development of modern specialty industries such as fruit tree planting, which benefits both the economy and ecology. These industries offer an increasing number of jobs and other opportunities for villagers to earn a better living in Xihaigu.

People power

A number of government officials, teachers, medical workers, professors, experts, entrepreneurs and volunteers from Fujian have contributed to Ningxia's poverty reduction.

In 1998, experts on mushroom growing and technicians from Fujian came to Minning, which at that time was a fledgling village. They offered guidance to villagers on building hundreds of mushroom greenhouses from scratch.

Other commercial crops were planted following similar paths. With the development of a planting and breeding industry and an increasing number of migrants, more villages sprouted up in the area. In 2001, Minning Town was established. Now it has six administrative villages under it.

The wine industry is also a pillar industry in Minning, which has suitable natural conditions. At the end of the 1990s, just a few farmers planted grapes and made wine in family workshops, not on a large scale. Later on, facilitated by the Fujian-Ningxia partnership, the industry gained a bigger scale thanks to the push of businesspeople from Fujian.

Chen Deqi is one of them. He decided to set up vineyards in Minning after he visited the place in 2007. His investment has turned into nearly 5 million grape vines and over 3,000 hectares of an organic grape industrial park. His vineyard employs more than 3,000 villagers.

"This is just the beginning," Chen said, adding that he plans to double the scale of his industrial park to provide more than 10,000 jobs for local residents.

By the end of 2019, 5,700 Fujian companies had participated in the industrial development of Ningxia, covering more than 20 industries and contributing tens of billions of yuan in annual output.

The Fujian-Ningxia partnership extends beyond economic development. In the past 24 years, more than 1,000 teachers and over 260 college students from Fujian went to middle and primary schools in Ningxia to teach students. Fujian has built 236 schools in Ningxia and assisted more than 90,000 poor students.

The Fujian Provincial Government also participated in over 300 medical and health programs in Ningxia including building maternal and child care service centers and healthcare training institutions.

Over 180 Fujian officials have taken temporary positions in Ningxia working to reduce poverty.

Liang Jiyu, head of the Poverty Alleviation Office in Ningxia, said the people from Fujian working in Ningxia not only bring business, capital and technology, but also advanced concepts for development and rich experience.

"Both Fujian and Ningxia, though thousands of miles away, share the same wish for development. The two regions will enhance mutual assistance and mutual learning, and start a new round of collaboration," said Hu Jiayue, an official of the Poverty Alleviation Office of Putian in Fujian.

Xinhua News
China's Major Winemaking Region Dreams Big

xinhuanet.com/english/2020-10/24/c_139464842.htm

24 October 2020

An international wine expo that ended Friday once again put China's Ningxia Hui Autonomous Region under the spotlight.

The two-day event, namely the ninth Ningxia International Wine Expo at Helan Mountains' Eastern Foot, attracted hundreds of experts from 96 countries and regions to attend online and offline to share opinions on the development of the wine industry.

Ningxia has made great achievements with its product quality improved and brand polished under preferential policies for the wine industry, said Regina Vanderlinde, president of the Paris—based International Organisation of Vine and

Wine.

Over the years, Ningxia has been rapidly transforming itself into a major winemaking region in China. It currently has 32,800 hectares of grape plantation, generating an annual output of 130 million bottles of wine and a comprehensive output value of 26.1 billion yuan (about 3.9 billion U.S. dollars).

Ambition runs high as Ningxia plans to double the grape-growing area by 2025, with an annual production of 300 million bottles of wine.

Vineyards in Gobi Desert

With abundant sunshine, rich irrigation and an appropriate climate, Ningxia's vast barren Gobi Desert is deemed the "golden zone" for growing wine grapes.

Dating back to 1984, Yu Huiming, along with seven young winemakers, spent four months making wine in more than 100 pickle jars.

At that time, many Chinese wineries closed down due to a lack of industry standards, and few customers knew about wine. The winery Yu worked for, Ningxia's first, also underwent a rough time. "Workers couldn't get paid timely, and six of my colleagues left for a better living," Yu recalled.

However, Yu chose to stick to the industry.

In the 1990s, a foreign company bought the winery's overstocked wine and then sold the wine with the company's own label at 268 yuan per bottle.

"I realized that our wine has a market, but we needed our own brand to tap into the market," said Yu.

In 2011, Ningxia launched a basket of favorable policies to regulate the wine industry and encourage wineries to produce high-quality and branded wine.

Since then, the preferential policies have attracted a large number of chateaus from home and abroad to establish wineries in the region. As a result, batches of high-quality Ningxia wine flowed into the market.

So far, more than 1,000 varieties of wine from 50 local chateaus have won medals in top wine contests around the world. Ningxia's wine has been exported to more than 20 countries and regions.

Dreamland for winemakers

Liao Zusong, 35, chief winemaker of the Xige Estate, a chateau in Ningxia, returned from an Australian wine farm in 2014. He participated in the whole winemaking process—from the construction of Xige's vineyard and winery as well as the design of winemaking techniques.

"I want to be a winemaker who knows the whole production industry chain from grape growing to vinification," Liao said.

Xige now boasts of over 1,333 hectares of vineyards with an annual output of more than 1 million bottles. "I believe we can produce the best wine for customers across the country and even the world," said Liao.

Meanwhile, to polish local wine brands and improve winemaking skills, Ningxia has also invited winemakers from world-famous wine regions. So far, 60 winemakers from 23 countries and regions have come to Ningxia.

Booming wine tourism

Zhao Liang, a tourist and also a wine lover from Beijing, drove all the way to Xige Estate with his friends. "I have been to many foreign wine farms, but I'm still surprised by the taste of Ningxia wine and the advanced production line of local wine farms," he said.

"The chateau is also close to many tourist destinations. It's especially suitable for developing wine tourism," said Zhang Yanzhi, founder of Xige Estate.

Ningxia has been tapping the potential of the wine tourism market. Currently, Ningxia chateaus receive more than 600,000 tourists a year.

"The wine-driven tourism plays a key role in promoting the region's tourism industry and also fuels wine consumption," according to Zhao Shihua, an official with the grape industrial park of the eastern foot of Helan Mountains.

South China Morning Post
Why China's Biodynamic Wines are Gaining a Cult Following

scmp.com/magazines/style/luxury/article/3044436/why-chinas-biodynamic-wines-are-gaining-cult-following-home

By Cybil Huichen Chou
3 January 2020

Winemakers imbibing the Chinese wisdom of "24 solar terms", an ancient calendar that governs agricultural arrangements even today, find it a good reference for practising biodynamic farming, as both systems have much in common.

"A Chinese book published in the sixth century really comes in handy for deciding when to sow, prune, harvest in vineyards or taste wines, as similarly instructed in the biodynamic calendar. Follow these principles and the vineyards can well be managed in good order," explains Silver Heights' [Emma] Gao.

As more wineries in Ningxia convert to organic farming, and biodynamic wines gain popularity, Domaine des Arômes has developed a Chinese cult following, mainly well-educated, middle-aged male consumers.(This is an excerpt. Use the url at the start for the full article.)

The Globe and Mail
The First Real Taste of Chinese Wine Shows Tremendous Potential

theglobeandmail.com/life/food-and-wine/article-the-first-real-taste-of-chinese-wine-shows-tremendous-potential/

By Christopher Waters
24 April 2020

An LCBO release of four wines from Chateau Changyu Moser XV, a boutique winery operating in the foothills of the Helan Mountain in Ningxia in northern China, is one of the first real glimpses of that country's wine boom.

[Ningxia's] focus on quality has many experts banking it is the region that will put Chinese wine on the map. Parallels have been drawn to the Napa Valley, which only makes 4 per cent of California's wine, yet is widely viewed as the Golden State's best. (This is an excerpt. Use the url at the start for the full article.)

Forbes
From Competition to Cantina: Concours Mondial de Bruxelles Opens Its First Wine Bar in Mexico City

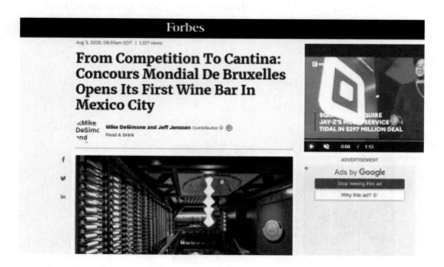

forbes.com/sites/theworldwineguys/2020/08/03/from-competition-to-cantina-concours-mondial-de-bruxelles-opens-its-first-wine-bar-in-mexico-city/

By Mike DeSimone & Jeff Jenssen
3 August 2020

Wine Bar by CMB is conceived as a space where wine and spirits producers from around the world can engage with wine and spirits enthusiasts as well as industry members such as sommeliers, distributors and importers…

 Among its various areas is the Ningxia Wine Pavilion, the first of its kind in Latin America, which is decorated in authentic Chinese style and offers award-winning wines from Ningxia, China. (This is an excerpt. Use the url at the start for the full article.)

China Daily
Ningxia Named Among World's 10 Most Promising Wine Destinations

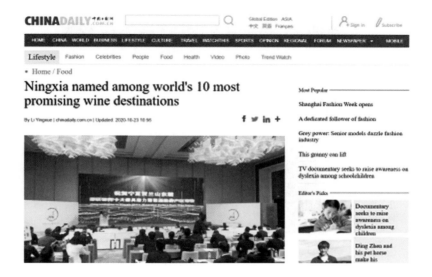

global.chinadaily.com.cn/a/202010/23/WS5f924640a31024ad0ba80897.html

By Li Yingxue
10 October 2020

The list of the World's Top 10 Most Promising Wine Tourism Destinations was revealed with Ningxia among the sought–after destinations.

The other destinations include Chile's Maipo, New Zealand's Wairarapa, Quebec in Canada and Yamanashi in Japan. (This is an excerpt. Use the url at the start for the full article.)

BBC News
China's Drinkers Develop Taste for Home-grown Wines

bbc.com/news/business-55226201

By Tim McDonald
17 December 2020

[Emma] Gao, who learned winemaking in France's Bordeaux region, agrees that if there's a shift towards Chinese wine, it is because of higher standards, and producers who know their market.

"I believe the quality of Chinese wine keeps improving," she says. "And this has been matched by a new generation of wine lovers that are more adventurous, proud of their Chinese heritage." (This is an excerpt. Use the url at the start for the full article.)

2021

La Revue du Vin de France
Ningxia, le Vin à la Conquête du Désert

larvf.com/,vins-chine-grace-vineyard-ningxia-vignes-wineries,13186,4246418.asp

By Jérôme Baudouin
2 February 2021

JOUR 5. Notre envoyé spécial, Jérôme Baudouin, s'est rendu dans la région chinoise de Ningxia, où le désert a vu naître la vigne en 10 ans!

On quitte Grace Vineyard pour l'aéroport de Taiyuan, à une heure de là. En route j'admire le paysage aride. Et je me rends compte combien la vigne est invisible. Elle était visible autour de Grace Vineyard. Des centaines d'hectares plantés sur un plateau limoneux. Mais depuis, plus rien. En fait, ce domaine n'est qu'un îlot posé au cœur du Shanxi. Folie de Zhi Qiang Chen, ce richissime homme d'affaires de la région qui a fait fortune dans le maïset et qui a voulu créer son vignoble.

Dans le Ningxia, notre destination, la situation est toute autre. Nous nous enfonçons plus vers l'ouest, au-delà du désert des Ordos que nous survolons. C'est la région du peuple Hui, à dominante musulmane, où la vigne est cultivée depuis des siècles. Une petite province coincée entre une steppe désertique sculptée par le Fleuve Jaune, situé au sud-est et les montagnes de Helan Shan, au nord-ouest. Elles-mêmes sont une barrière naturelle au désert de Tenggeli qui s'étend quant à lui plus au nord, au-delà des montagnes d'Helan Shan. Il signale le début de la Mongolie Intérieure. Seul le Fleuve Jaune rend ici les terres cultivables.

Dès notre première visite dans le vignoble, à Qingtongxia, je comprends comment la vigne est cultivée ainsi que toutes les cultures. Les rangs de vignes cotoient des rizières et les champs de maïs. En fait, un gigantesque réseau d'irrigation a été mis en place pour capter l'eau du Fleuve Jaune. Un agronome du ministère de l'agriculture de la province de Ningxia nous explique combien ce choix est stratégique. "Il faut imaginer qu'ici, il y a dix ans, il n'y avait rien, que du sable. Nous avons mis en culture des dizaines de milliers d'hectares et développé une activité agricole riche de main d'œuvre" détaille le fonctionnaire, face à des immeubles en construction.

Et la production de vin est aujourd'hui considérée comme un secteur stratégique, à très haute valeur ajoutée, d'où cet engouement autant de la part des investisseurs privés, que du gouvernement qui n'hésite pas à mettre en place des systèmes d'aide au développement. Plus tard, on nous montre des photos où une dizaine de bulldozers en ligne avance dans un décor désertique pour préparer la terre, creuser les sillons et les rigoles d'irrigation. Le combat semble déjà gagné. L'eau coule à flots depuis ce fleuve intarissable alimenté par les grands glaciers de l'Himalaya. Le Ningxia compte déjà 24,600 ha de vignes destinées à la production de vin et une quarantaine de wineries de différentes tailles. Ningxia est aujourd'hui considérée comme la première région de production viticole de qualité de Chine et en mars 2012 elle a fait son entrée en tant qu'observatrice à l'OIV (Organisation internationale de la vigne et vin). Ningxia n'est plus un désert. (This is an excerpt. Use the url at the start for the full article.)

The Irish Times
Enter the Dragon: Chinese Wines Embodying Lessons from the West

irishtimes.com/life-and-style/food-and-drink/drink/enter-the-dragon-chinese-wines-embodying-lessons-from-the-west-1.4469289

By John Wilson
6 February 2021

One of my more interesting Zoom tastings last year was with Austrian wine producer Lenz Moser. We weren't tasting Austrian wine, though: instead it was Chinese wine. Moser has decided to focus on Chateau Changyu Moser XV in Ningxia in China, where he makes the wine as well as looking after global sales.

"Twenty-fifteen was the breakthrough vintage for us," he says. "We harvested two weeks later than usual, because the moon festival was early, and all of the pickers disappeared to celebrate. The wine was instantly better, with more alcohol and ripe tannins—previously it had been 12.5 per cent and herbaceous like a Loire Cabernet Franc." (This is an excerpt. Use the url at the start for the full article.)

jamessuckling.com
2020 Ningxia Report: Fast Money vs Age-worthy Quality

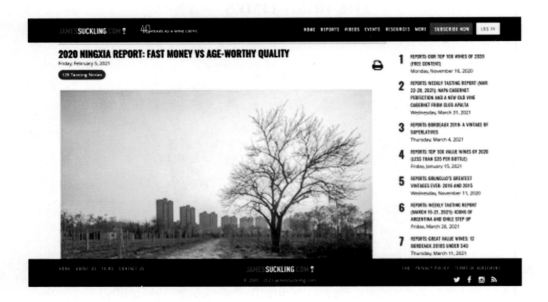

jamessuckling.com/wine-tasting-reports/2020-ningxia-report/

By Zekun Shuai
5 February 2021

Located in the central northwest of China, Ningxia Hui Autonomous Region has been developing its wine industry with "China speed". Only five or six years ago there were far fewer wineries in operation, and those that were making wines were still learning the intricacies of producing quality wines.

Today the number of wineries in Ningxia, which is 1,100 kilometers from Beijing and situated on an open plateau 1,000 meters above sea level, has mushroomed to nearly 100, and there are at least another 100 waiting for government approval to start producing wines. Some are modest, some lavish, some bizarrely Western, and some ingeniously modern and unique. And they are no longer a glamorous facade—the wine that comes out of many of these wineries is now the real deal. Yes there are still some problems with formula winemaking, but the road to great wine has been paved.

This is our first close-up review on Ningxia, which features around 130 wines from the region, including a few tasted in the wineries: Helan Qingxue, Domaine des Aromes, Legacy Peak, Lilan Winery, Chateau Mihope, Xige Estate and Silver Heights.

Most producers sent samples on request (including a few purchased samples) as their wines were already on our radar. The results delivered just a few disappointments despite a relatively high number of corked wines—six bottles in total. Considering that almost one-third of the wines now use DIAM corks, that cork taint rate is very high, and this is a noticeable issue here. Winemaker Gu Jiawei from the Ruwen winery explained that the natural corks exported to China from Portugal and Spain often turn out to be their lower-quality closures, which is why many serious producers have turned to DIAM corks over the last few years. Ruwen is now experimenting with different stoppers, including one made of bamboo that also blocks TCA.

Of the 130 samples tasted, many found a comfort zone within a range of 88-90 points, but only 13 wines made it to 93 points or above. These were wines with real character instead of just being flattering and generous (especially in sweetness), which is a commonly found quality from the reds in Ningxia.

Some would say the "sweet fruit" is what Ningxia red wine is about, which has to do with the climate and terroir. Certainly, the climate gives a rich, fruity context to varietals like cabernet and merlot, but I couldn't help wondering if this was also just an excuse for not trying hard enough or for picking late to fit the expectations of the market, something that will have an even bigger impact on the style than terroir and climate. After all, the top wines in this report have little problem with that cloying character. Nevertheless, after tasting 130 wines Ningxia has lived up to the expectations given the numerous medals the region receives year after year. But there is still a lot of work to do if wineries want to be rigorous and unremitting with quality, and aspire for more in the future.

Today, Ningxia makes solid, rich, yet heady reds with that "flattering" sweetness. Undoubtedly, the fruit generosity enhanced by a supple mouthfeel thanks to high alcohol (some over 15.5%) and glycerol is very New World-like, and some also left a bit of residual sugar around 4-5g/L to make it even more appealing to the inexperienced, new consumers. As cabernet sauvignon struggles to ripen here (when it does, without going overripe, it makes beautiful wines), many producers started to harvest late to make sure the wine does not have an unpleasantly lifted bell-pepper character, something common 10 years ago. "Most Chinese consumers don't like acidity and tannin" was a common comment when I spoke to producers, and being Chinese myself, I know this is the case for most consumers who just started drinking wine.

But to keep the balance in check and stabilize the wine, many producers tend to acidify the wines in hot vintages. It probably explains why some wines tasted rich, even jammy, yet have a tart attack on the palate instead of unleashing the juicy character from the fruit.

As this style appears to be well received in the market, some producers follow this formula and neglect styles that are more authentic and complex. But the top producers in Ningxia understand this issue, and tend to limit this category of simple, fruity wines to their entry-level labels to attract new consumers. Some also avoid this "sugar coating" of their wines. Zhang Jing is one of them, the co-founder and winemaker of the leading estate Helan Qingxue, who told me that she wanted her wine to have real depth.

"I want my wines to show some depth at youth, which helps them age. We've also cut acidification since 2017, as we feel it is best to keep the acid as it is. As we always pick early, the acid is not an issue for us," said Jing. Both red and white, her wines already show a restrained style with freshness and subtlety, rendering a classy and precise feel with a reserved modernness to their Old-World sensibility.

Also made by a female winemaker, Silver Heights produced an array of wines that showed the winemaker's intuition. "Precision" is probably not the best word for Emma Gao, the winemaker who, like her father Mr. Gao Lin, has a vision for wine and the region. Emma Gao's wines are full of nerves with an authentic and unfettered personality, a style that I would describe as a combination of Ningxia's true generous fruit with her winemaking know-how. "It is like our Ningxia people," said Emma in a soft voice, referring to the down-to-earth, hospitable personality of the local people who might get offended if their guests won't stay for the local roast lamb.

Today, Emma also wants to make wines as authentic and natural as possible. To pay more respect to nature and to make vines healthier with a better "immune" system, her vineyards were converted to biodynamic farming in 2017. "We have worked hard to understand our land better so the wines could have soul and less manipulation," said Emma, and their new line Jiayuan (Homeland) made with wild yeast and a low dosage of sulfur reveals her aspiration to produce wines that highlight the typicity of the grape variety as well as the place.

When to start the harvest is a crucial but tricky decision in Ningxia, as the growing season isn't long enough despite a vast diurnal temperature difference and plenty of sunlight which gives a high amount of sugar/alcohol. So, it can be a bit hit or miss. Either you have lifted greenness if picked too early without attractive phenolic ripeness, or you have overripe fruit and very high potential alcohol with low acidity if picked too late. But if cabernet sauvignon has found a sweet spot for both ripeness and freshness, these wines from Ningxia are not only rich and welcoming but also classy, fresh, and refined. Our top cabernet wines in the list showed this character with a sophisticated application of oak barrels. In fact, many top wines come from barrel selections, such as the magnum Silver Heights Emma's Reserve and Helan Qingxue's Special Reserve 2016.

Among the top wines that scored 93 to 95 pts, most are cabernet sauvignon or Bordeaux blends, and many come from my four top producers, which are

Helan Qingxue, Silver Heights, Kanaan Winery, and the fledging Jade Vineyard, whose wines blew me away when I first tasted them in 2016. Today, I believe these four producers are leading the game in Ningxia, and there is a fun fact with these producers: they are all led by female winemakers/owners, and two of them (Jade Vineyard and Kanaan) are consulted by Zhou Shuzhen, an experienced and well-respected independent winemaker today in Ningxia who also consults for wineries like Li's, Legacy Peak, and Copower Jade.

Jade Vineyard Ningxia Messenger Reserve 2016 (95 points) was my top wine among the 130 bottles. It shows how good Ningxia's cabernet sauvignon can be when the producer gets it right. The successful 2016 vintage also helped forge a compelling interplay between ripeness and complexity, delivering a super long and hedonistic wine with a dense, serious yet superbly polished tannin structure. Helan Qingxue, Kanaan Winery, and Silver Heights all delivered a few cabernet sauvignons/Bordeaux blends that scored 94 points, showing cabernet sauvignon's potential in Ningxia when it is perfectly ripe, and not over the top.

What surprised me in my tastings were three pinot noirs and a few other grapes, including syrah, viognier, and even malbec. For pinot, Deng Zhongxiang, an aspiring winemaker educated in Burgundy who consults for several wineries including Lansai and Domaine Charme, told me that it is indeed a challenge to make excellent pinot noirs from Ningxia, but not impossible.

"Ningxia can be too hot for pinot noir with a shorter growing season, but despite these odds, I want to show the world it is not impossible, and by harvesting early, we don't lose the natural acidity and the subtle characters of pinot noir," said Deng. His Lansai Ningxia Yinchuan Hong Pinot Noir 2017 (92 points) was an excellent example of pinot showing ethereal fruit, elegance and details, even if it is framed in a warmer context. Silver Heights Ningxia Jiayuan Pinot Noir 2017 (93 pts) was one of the highlights that showed the cloudy, sous-bois character despite some spicy new oak, followed by a firm backbone of acidity on the palate in a linear, almost demandingly way. Helan Qingxue Ningxia Jiabeilan Baby Feet Pinot Noir 2017 (92 pts) was a little more compact and richer, underscoring the depth of fruit and color, with a floral and slightly stemmy character bringing out freshness in a broad and cohesive palate.

Together with Chateau Chanson, whose 2017 pinot noir we tasted earlier, these producers might turn your perception of Ningxia around, given their record of producing excellent wines from one of the finest yet most finicky grapes in the world. The best pinots that I came across certainly cleared my doubt about Ningxia's future possibilities. The question now isn't if Ningxia has the potential for more elegant grapes but rather if the producers/winemakers have the willingness and commitment to nurture them; whether they dare travel on a difficult and offbeat road, such as producing an excellent pinot noir.

My other highlights from Ningxia were dotted with a few grape varieties. We already know that marselan is strong in China, with the best often showing a deep, brooding color, with tarry, spicy dark berry fruit, and blue flower characters. Domaine Franco-Chinois from Hebei makes one of the most essential wines out of it. In Ningxia, the popularity of marselan is also growing, and some exude an exotic black pepper, lavender, and almost dried lychee character, although lesser examples went a little too far for ripeness, letting too much dried fruit, alcohol or jamminess jump out. The best marselans from this report came from Jade Vineyard and Domaine Charme: both are barrel samples from 2019, an outstanding vintage for Ningxia.

Besides marselan, syrah is on the rise, too. Chateau Rong Yuan Mei is another winery consulted by Deng Zhongxiang and makes an impressive syrah and an excellent malbec. The latter might become popular in the region considering its high altitude and dry climate that reminds people of Mendoza. The young winemaker Liang Ning showed how enticing syrah can be in Ningxia with his Sweet Dew Vineyard Ningxia Syrah Limited Edition 2017 (93 pts), a very promising bottle that knits up intensity and balance within its focused blue fruit, spicy oak, and a drizzle of tangy black pepper spices. Li's, a winery with a real estate background, is another producer to watch for syrah.

While white wine is not on every producer's radar as China is still more of a red wine market, if wineries want to produce one, chardonnay is the most-picked grape. That said, Domaine Charme and Chateau Mihope have made some eye-opening viogniers showing the floral, aromatic quality and the stature of the variety. Try Domaine Charme Ningxia Viognier 2019 (93pts) for its zingy nose full of lemongrass, citrus, and peachy aromatics that are streamlined on the palate with a vibrant streak of acidity, which leads the fruit to a more subtle, chalky finish.

These wines also showed how younger winemakers that are gradually taking things beyond what they learned from their mentors, and who have dotted Ningxia's cabernet canvas with new varieties, ideas and interest, are injecting vigor to a region that is not much older than themselves.

As wine blooms, Ningxia now plans to get wineries more involved in tourism. Producers' ability to receive tourists is also taken into account in a young classification that the government has run since 2013. And if you are planning a trip to Ningxia some day, there is one producer that you should not miss. An hour's drive from Yinchuan to the south, you will find the brand-new Xige Estate, a hidden gem in the Gobi Desert, with its stunning architecture and convincing wines. Today, this ambitious and modern winery with an avant-garde design holds some 1,300 hectares of vines aiming to produce quality wines in quantity. Among them, Xige Estate Ningxia Jade Dove Cabernet Gernisht Single Vineyard 2018 (93 pts) shows how compelling cabernet Gernisht can be when it fully ripens in Ningxia, as, for a long time, the variety was considered a no-go with its strong herbal, leafy characters and problems with viruses. Li's also produced an outstanding example with its Li's Ningxia Family Selections "Justice" Cabernet Gernisht 2017 which received 92 points, a delicious wine showing varietal typicity underlined by its dried herbs and spicy fruit that pinpointed to the finish. It is also one of the best values in this report, as I bought it at a price of RMB 199($30), which is extraordinary for Ningxia with most of its quality wines sold for much more, many around RMB 300-600 ($45-$85) at least.

"Cabernet Gernisht needs to be picked late enough as it is difficult to ripen," said Liao Zusong, a young and extremely committed winemaker of Xige Estate, who previously worked under Bass Phillip and now oversees the winemaking team along with the owner Zhang Yanzhi, a Bordeaux-trained oenologist who started this aspiring project after his successful business with Penfolds Max's. "We have around 1,000 hectares of vines over 23 years old," said Liao, which arguably gives them an edge in more concentrated fruit.

In Ningxia, a 23-year-old vine is already considered old. Legacy Peak, a boutique winery located within the heritage park of Xixia King's tomb, also owns hectares of these much-coveted old vines planted in 1997. Their vineyards are just miles away from the Eastern foothills of Helan Mountain on a higher spot in the region above 1,200m in altitude.

Another problem for Ningxia, indeed for most vineyards in northern China, is that when the vines get old enough, vine burying, a costly vineyard management that is still the most effective protection for vines to survive mercury, becomes a challenge. Old vines are more likely to be snapped in this rather brutal practice that usually starts in early November before the soil freezes, even if the vines have been trained to facilitate this process.

"It can be a problem for vines on their way to 30 years old when the trunk is too hard and thick to endure the burying," said Liang Ning, winemaker of Sweet Dew Vineyard.

However, Liao Zusong of Xige Estate told me they were planning ahead. "We have nurtured new shoots for the old vines that will become the trunk in the future, so we can cut off the old trunks, if necessary, but keep their roots." Still, the life expectancy of vines is not predictable for those regions that need to bury the vines in the northern part of China. Having said that, in a freezing morning after snow, when your eyes lay on the lonesome stony poles standing like soldiers that guard an "empty" vineyard blanketed with soil and snow, there is also something special to this, a real beauty of silence and solemnity.

Ningxia's First Growth in Labor

As an embryonic wine region, Ningxia could not have accomplished so much in such a short period (10-15 years) without the help of the local government, whose involvement also encouraged investment from other industries, from energy enterprises and home appliance tycoons to real estate developers. This brings us back to Ningxia's Grand Cru Classification, an official ranking since 2013 which now gathers 37 wineries in its 2019 updated list. While speediness is not news for China, the government wants to focus on long-term development by encouraging producers to make better wines, which was why the grand cru system, a classification learnt from Bordeaux, was created. However, a few prominent names are missing.

Most winemakers I talked with believe the grand cru classification is applaudable for an emerging wine region like Ningxia; at least the pros outweigh the cons. Yet other voices also exist. "It is premature and way too early to rush into a ranking of producers now. Compared to Bordeaux, Ningxia is a small baby. What if a grand cru winery selected today didn't understand their soil and mispositioned itself in the market, and it had to shut down one day? It's a good thing to do for sure, but we cannot rush it," said Gao Lin, as we sipped some hot tea in his Silver Heights' cultural center, which was still under construction. As a leading boutique estate in Ningxia and a family business, Silver Heights is not on the grand cru list, and it's possible it will remain separate for a long time. Curiously, another extremely serious producer, Jade Vineyard, was not on the 2019 list either. Perhaps, as Gao Lin said, it is not the time yet.

"There should be some self-examination, and different voices should be heard, or at least we need to calm down and think about it before we get heady and everyone jumps in with both feet," added Gao Lin.

Mr. Gao Lin had a point. Winemaking in Ningxia has only just left the starting line and there's a marathon ahead. Many winemakers also told me that only a few wineries in Ningxia are profitable, and many still struggled to survive.

Deng Zhongxiang, however, has few misgivings and said he is a "wholehearted supporter" of the classification. "With so many producers and brands now, the grand cru classification is like a Ningxia wine guide, and it certainly helps consumers pick out better wines from the myriad of producers and brands. Of course, there are imperfections in this system that have drawn a lot of criticism, but it is a good start, and isn't almost any greatness started out amid criticism and skepticism?" laughed Deng.

The evaluation of Ningxia's grand cru wineries is based on the government's official document (in Chinese) of "Interim Measures for the Evaluation and Management of Classified Wineries in the Eastern Foothills of Helan Mountains Wine Region," which stipulates the rules. Wineries are subject to assessment every two years in different aspects such as vineyards (wineries must own vines over five years old), yield, price, sensory evaluation through a jury tasting, medals and the market value of the brand (including the reputation of the winemaker), wine tourism reception ability, etc. Each winery eventually receives an aggregated score after which it may receive promotion/demotion. New candidates could be brought in and existing ones could be removed, too. "It is a pretty dynamic list," commented Deng.

In a webinar last year, I remember Emma Gao said that Silver Heights is still a family business, and they don't have the reception facilities required to be considered for the list. Tiny, but serious and quality-driven producers such as Domaine des Aromes might also get sieved out in this "Hall of Fame". More importantly, it comes down to each producer's drive and their willingness to deal with the administrative side of honors and recognitions, things that might not happen in a vineyard or cellar, but instead in an office or in front of a computer filling out application forms. After all, producers have to apply for the entry in the first place.

During my visit to Chateau Mihope, a property owned by the Chinese electronic appliance tycoon Midea Group, its winemaker Zhou Xing also shared some of his insights on the classification.

"The focus of each growth is slightly different. For the 5th growth, the key is the vineyard, which requires wineries to have their own vineyards of vines over five years old. The 4th growth also considers vinification upon viticulture. For the 3rd growth, you also have to include the market, price, and reception facilities. For the 2nd growth, brand image and reputation are considered, too," said Zhou.

With the highest aggregated score of 150.1 points among the 4th growths, Chateau Mihope seems to be a strong candidate for a 3rd growth promotion. Currently, three producers sit at the top of the pyramid (the 2nd growth): Chateau Yuanshi (180.92pts), Helan Qingxue (180.42pts), and Chateau Bacchus (174.73pts). The next update of Ningxia's grand cru list is expected to come out soon this year, and for the first time, Ningxia may have a chance, if they will, to roll out the red carpet for its first 1st growth winery.

The touchstone vintages

My trip to Ningxia in November was certainly productive and rewarding. There are so many stories to account from each producer and we can't cover them all in a single report. Yet, there is no better narrator than a bottle of wine itself which recalls for us the four seasons in a single year. For Ningxia, there were, of course, some ups and downs for vintages, but the top producers have worked hard to stay consistent.

"2016 should be more unanimous among producers marked by its good ripeness. But I think vintages like 2015, 2017 and especially the lesser 2018 are the better touchstones if you really want to see what the winery is capable of," said Zhang Jing of Helan Qingxue.

Compared with the hot 2017, 2018 was considered a very challenging vintage for Ningxia underscored by frost and rains. "It is one of the coolest and wettest vintages with an average rainfall around 490mm compared to some 270mm in 2017 and the excellent 2019, which only recorded roughly 170mm of rain," said Zhou Xing of Mihope. The average amount of rain in this dusty Gobi

Desert plain elevated to over 1,000m is usually less than 200mm a year, but much of it falls in the summer. Irrigation is a must.

"2019 would be an outstanding vintage for Ningxia," said Liu Jian Jun, who still has to rely on purchased grapes to make his wines. His Lingering Cloud Ningxia Red Beard 2017 (93 pts) was one of our highlights last year from China. In 2018, Liu did not want to compromise, and downgraded all his wine and sold it under his second label Moby Dick. Zhang Jing of Helan Qingxue also told me that they had to discard almost half of the red's crop in 2018 to ensure the cabernet sauvignon quality.

From the tasting point of view, wines that succeeded in 2018 were elegant and balanced, even if they lacked a bit of ambition, which was the case for some lesser examples with dilute mid-palates overwhelmed by broad-shouldered tannins. Producers needed to be extra-careful with the tannin extraction based on the less natural concentration of 2018. But making wines with more finesse is possible, and the best producers know when to let go of some concentration in exchange of freshness and crunchier fruit with more natural acidity than those of 2017, a vintage marked by its richness and high alcohol for many reds—around 15.5% or more.

That said, the top producers made balanced and soulful wines from both 2017 and 2018, with a healthy fruit concentration, balance and tannin quality. As we talk about climate, vintage and terroir, and all the miracles and efforts happening in the vineyards, it finally comes down to the producers to define the vintage and tune in to their "house style," if there is one. So, should we let the climate and terroir take full ownership of the style? Probably not.

Noticeably, those that always produce jammy, heavy and high-octane red wines follow a recipe to flatter the market. Instead of subtlety and nuanced complexity, they are fond of "cutting corners," delivering flavors inexperienced consumers can easily grasp. After all, China, in general, is not yet a mature wine market, and wine appreciation has just started. The WSET program has helped, but at the time of writing has had to suspend all its classes in China.

For Ningxia, the strength of very sweet fruit, supple, round tannins and moderate acidity, could captivate a following in the domestic market. There is nothing wrong with it, as wine is essentially a product to be sold to consumers, and few producers can afford the time and the money to educate the masses about their demanding wines with complexity, depth, finesse, and structure. And of course, producers have to work extremely hard to make such wines. So, it all comes down to who are the consumers that wineries are targeting now.

"Many of us still don't know that clearly, and that's a problem for Ningxia," said Wang Fang, the owner of Kanaan Winery, during an interview last year. As wines get better and better, it seems now it is time for producers to find out more about the market and the people who drink their wines.

Ningxia has a real chance now in the next 5-10 years, to flourish in the world's second-biggest wine market, according to IWSR and Vinexpo predictions. Now, it is up to producers to decide if they want to deliver the fruity, overly generous style of wines that many think the mainstream likes and make some fast money, or to make even better wines with freshness, depth, nuanced complexity and real structure, selling for a reasonable price, in a way that could engage consumers for the long term. It could be a more challenging road for smaller and less resourceful producers for sure, but as Chinese wine is coming of age and Ningxia is center stage, this mindset will separate the best producers from the good ones.